SpringerBriefs in Fire

Series Editor

James A. Milke

For further volumes:
http://www.springer.com/series/10476

Brian Meacham · Brandon Poole
Juan Echeverria · Raymond Cheng

Fire Safety Challenges of Green Buildings

 Springer

Brian Meacham
Brandon Poole
Juan Echeverria
Raymond Cheng
Worcester Polytechnic Institute
Worcester, MA
USA

ISSN 2193-6595   ISSN 2193-6609   (electronic)
ISBN 978-1-4614-8141-6   ISBN 978-1-4614-8142-3   (eBook)
DOI 10.1007/978-1-4614-8142-3
Springer New York Heidelberg Dordrecht London

Library of Congress Control Number: 2013942653

Printed on acid-free paper

Springer is part of Springer Science+Business Media (www.springer.com)

Foreword

Many new commercial facilities are being designed and constructed with an objective of achieving a "green building" certification. There are many sustainable building features and products that singly or together may have an impact on fire safety unless there is a design approach which mitigates those effects. The Foundation commissioned this study to develop a baseline of information on the intersection of "green building" design and fire safety and to identify gaps and specific research needs associated with understanding and addressing fire risk and hazards with green building design.

Keywords Sustainability • Green building • Building design • Fire protection

The Foundation acknowledges the contributions of the following individuals and organizations to this project:

Project Panel

David Barber, Arup
Anthony Hamins, NIST Engineering Laboratory
Debbie Smith, BRE Global, UK
Craig Hofmeister, RJA Group
Tracy Golinveaux, NFPA codes and standards staff liaison

Sponsors

Zurich Insurance
FM Global
Liberty Mutual
CNA Insurance
Travelers Insurance
Tokio Marine
XL Gaps

Preface

A global literature review was undertaken to (a) identify actual incidents of fires in green buildings or involving green building elements, (b) identify issues with green building elements or features which, without mitigating strategies, increase fire risk, decrease safety or decrease building performance in comparison with conventional construction, (c) identify reports, studies, and best practice cases which speak to the issue of addressing fire risk introduced by specific green building design elements, and (d) identify research studies in which building safety, life safety, and fire safety have been incorporated as an explicit element in green building indices. In addition, consideration was given to how one might express the level of increased risk or hazard, or decreased performance, associated with fire performance of green building features. Steps were also taken to identify gaps and specific research needs associated with understanding and addressing fire risk and hazards with green building design.

Outcomes of this effort include the following:

- There are currently no fire incident reporting systems in the United States or other countries surveyed which specifically collect and track data on fire incidents in green buildings or on items labeled as green building elements or features. Unless changes are made to reporting systems such as NFIRS, it will be difficult to track such fire incident data.
- Web searches and surveys have identified more than two dozen reported fire incidents that are related to green issues. Examples include fires associated with photovoltaic (PV) panels and roof materials, fire and safety hazards attributed to increased energy efficiency aims in residential buildings (primarily insulation related), fire involving insulating materials, fires associated with exterior cladding that contains combustible insulation materials or coatings, and fire performance of timber frame buildings with lightweight engineered lumber (LEL) components.
- Studies related to green building and fire issues produced by BRE, BRANZ, FMGlobal, and the NASFM have been identified. Research on specific building elements with green attributes, but not necessarily labeled as green, such as lightweight engineered lumber (LEL), has been identified at UL and NRC Canada.

- From the materials reviewed, a comprehensive list of green building site and design features/elements/attributes has been compiled. The list is titled "Table 4. Green Building/Site Elements and Attributes" and can be found in Chap. 3.
- From the materials reviewed, a list of fire-related hazards and risk factors, associated with green building elements, has been compiled. The list is titled "Table 5. Hazard, Risk, and Performance Attributes" and can be found in Chap. 4.
- Using Table 4 and Table 5, a set of matrices relating green attributes and potential fire hazards was developed. The matrix concept is illustrated in "Fig. 1. Matrix of Green Attributes and Potential Fire Hazards" and can be found in Chap. 5. The complete set of matrices is detailed in Appendix E.
- Using the matrices identified above, an approach for illustrating the relative fire risk or hazard, or decreased fire performance, associated with green building elements, was developed. The relative risk matrix is illustrated in "Fig. 2. Relative Fire Risk/Hazard Level of Green Attributes" and can be found in Chap. 5. A complete matrix, which is based on a qualitative expert judgment approach for illustrating relative risk levels, can be found in Appendix F.
- Potential mitigation strategies for addressing the relative increase in fire risk or hazard associated with the green building elements and features have been identified. These are presented at a basic level (e.g., provide automatic sprinkler protection). In many cases, adherence with existing test standards, codes, and related design guidelines associated with conventional construction will help mitigate potential increases in fire risk or hazard associated with green building elements. Approval or certification of products which meet loss prevention criteria, and are indicated as having some type of green attribute which might gain credit in a green certification scheme, have been identified within the FM Approvals system and UL Product Certification system.
- Review of a sample of green rating schemes for which data were readily available, including LEED (residential and retail), BREEAM (new buildings), GREEN MARK (residential and nonresidential), as well as the IgCC, indicates that fire safety objectives are not explicitly considered. However, as noted above, implementation of certain green features could have a negative impact on fire or life safety if not mitigated. A qualitative approach, using text and pictograms, was used to reflect areas of fire and life safety concern, as illustrated in "Fig. 4. Fire Hazards with Green Building Features" and "Fig. 5. Extract from GREEN MARK Assessment for Potential Unintended Fire Consequences", which can be found in Chap. 5. Detailed matrices of the assessment of the green rating schemes for potential unintended fire consequences can be found in Appendix G.
- It was determined that the green building rating scheme of the German Sustainable Building Council (DGNB) includes criteria for fire prevention (http://www.dgnb-system.de/dgnb-system/en/system/criteria/, accessed last on 10/29/12). Detail on the weights of fire prevention attributes relative to the green attributes was not able to be verified; however, it is understood that some credit is given for fire protection features such as smoke extract, automatic sprinklers, and structural fire protection.

- It was determined that BREEAM-in-USE (http://www.breeam.org/page.jsp?id=373, last accessed on 10/29/12), a recent BRE scheme to help building managers reduce the running costs and improve the environmental performance of existing buildings, incorporate fire risk reduction attributes. The fire risk reduction attributes related to such issues as whether a fire risk assessment has been conducted, are emergency plans in place, and so forth. No indication of consideration of fire protection systems was identified.

In order to fill gaps in knowledge to better address fire issues with green building features, further research is suggested in several areas.

- To address the lack of reported fire experience with green buildings and green building elements, especially in buildings which have a green rating or certification, a modification is required to fire incident data reporting systems as NFIRS. This could perhaps be done in collaboration with the USGBC and/or AIA, and parallel organizations in other countries. If this avenue is pursued, there will be challenges associated with how responding fire departments are able to identify LEED, BREEAM, or other such ratings for buildings. In cases where ratings or certifications are posted on buildings (e.g., LEED, Energy Star, BREEAM, or other), this information could be readily captured by fire fighters responding to fire incidents in the building. In cases where such ratings or certifications are not posted, inclusion of specific features such as 'double-wall façade' or 'LEL' might be added to the incident reporting system, or additional guidance can be provided to first responders in identifying green attributes of buildings.
- To address the lack of analysis on fire 'risk' associated with green building elements, it is suggested that a more extensive research project is needed to review existing studies and reports on fire performance of green building elements, even if not explicitly identified as such (e.g., LEL). Research is needed to (a) develop a clear set of comparative performance data between green and 'conventional' methods, (b), develop an approach to convert the relative performance data into relative risk or hazard measures, and (c) conduct a risk (or hazard) characterization and ranking exercise, with a representative group of stakeholders, to develop agreed risk/hazard/performance levels.
- To explore the extent to which current standard test methods are appropriate for evaluating both green and fire safety criteria, and result in adequate mitigation of fire risk/hazard concerns, investigation into level of fire performance delivered by current standard test methods and into the in situ fire performance of green building elements is recommended.
- To address the lack of published case studies in which increased fire risk or hazards associated with green building elements have been specifically addressed, groups such as SFPE, NFPA, AIA, and the USGBC can be encouraged to hold symposia on these topics and encourage publication of case studies in proceedings and associated journals. While such studies have been published, they mostly reflect 'issues or concerns' with green building features without significant quantification of impacts and formal risk analysis.

- To address the lack of studies which have investigated incorporating building safety, life safety, and fire safety as explicit elements in green building indices, joint research efforts between the FPRF and the USGBC, and other promulgators of such indices could be explored with the aim to incorporate fire and life safety objectives as fundamental elements in green rating schemes and codes.
- To facilitate better collection of relevant data on fire safety challenges with green buildings in the future, a fire and green building data repository could be established. This might build on an existing effort (e.g., http://www.firemarshals.org/programs/greenbuildingsandfiresafetyprojects.html) or be supported by the FPRF or other organizations.

Contents

1 Background and Introduction 1
  1.1 Problem Statement, Project Objectives and Tasks 1
    1.1.1 Problem Statement 1
    1.1.2 Project Objectives. 2
    1.1.3 Tasks. 2
  1.2 Additional Tasks. 2
  1.3 Research Direction and Observations 3

2 Information Search 5
  2.1 Representative Fire Incidents 5
  2.2 Selected Resources Related to Fire and Green
       Building Concerns 8
  2.3 International Survey and Responses. 11
  2.4 Review of Representative Green Building Rating Schemes
       for Fire Considerations. 11

3 Green Building/Site Elements and Attributes 15

**4 Attributes of Green Building or Site Which Could Impact Fire,
   Life Safety, Building or Fire Service Performance** 17

5 Hazard/Risk Assessment and Ranking 19
  5.1 Detailed Matrices of Green Building Elements/Features
       and Hazard/Risk Factors 21
  5.2 Tabular Representation of Potential Fire Hazards
       with Green Building Elements............................ 23
  5.3 Fire Hazards Associated with Green Rating Schemes and Codes ... 23

6 Summary and Conclusions. 29

Appendix A: Informational Resources 33

Appendix B: Representative Fire Issues and Mitigation Approaches 43

Appendix C: International Survey and Responses 57

**Appendix D: Detailed Matrices of Green Elements
                        and Potential Fire Hazards** 63

**Appendix E: Illustration of Relative Hazard Ranking
                        with Detailed Matrix** 67

**Appendix F: Tabular Representation of Fire Hazards
                        with Green Building Elements.** 71

**Appendix G: Assessment of LEED, BREEAM, GREENMARK
                        and the IgCC for Fire Safety Objectives** 81

References 83

Chapter 1
Background and Introduction

This report has been prepared for the Fire Protection Research Foundation (FPRF) as the deliverable for the project *Safety Challenges of Green Buildings*. Section 1.1 reflects the Problem Statement, Project Objectives and Tasks as specified in the request for proposal. Section 1.2 reflects additional tasks which were explored. Section 1.3 summarizes the direction taken based on research undertaken.

1.1 Problem Statement, Project Objectives and Tasks

1.1.1 Problem Statement

To enhance their market value, and improve the sustainability of operations, many new commercial and industrial facilities are being designed and constructed with the goal of achieving a "green" certification, the most common of which in the U.S. is the LEED certification by the U.S. Green Building Council. The International Code Council and other groups are promulgating "green" building codes. These changes to building design and materials are an opportunity for safety improvements but may also include building performance, fire and safety challenges that have unintended consequences for sustainability from property damage as well as life safety. An assessment of fire performance (among other considerations) of green buildings, and focused research on the primary changes affecting building performance, fire and safety, are required. Furthermore, a systematic method needs to be developed for implementation in the certification process that integrates the consideration of fire as well as other hazard risk factors as part of design performance metrics.

B. Meacham et al., *Fire Safety Challenges of Green Buildings*, SpringerBriefs in Fire, DOI: 10.1007/978-1-4614-8142-3_1, © Fire Protection Research Foundation 2012

1.1.2 Project Objectives

In addressing the problem as identified above, two objectives were provided:

1. Systematically document a set of green building design elements that increase safety hazards, and
2. Share best practices for hazard risk mitigation.

1.1.3 Tasks

To indicate progress on meeting the above objective, the following task was identified:

1. Conduct a global literature search to:

- Identify documented fire incidents in the built inventory of green buildings,
- Define a specific set of elements in green building design, including configuration and materials, which, without mitigating strategies, increase fire risk, decrease safety or decrease building performance in comparison with conventional construction,
- Identify and summarize existing best practice case studies in which the risk introduced by specific green building design elements has been explicitly addressed, and
- Compile research studies related to incorporating building safety, life safety and fire safety as an explicit element in green building indices, identifying gaps and specific needed research areas.

1.2 Additional Tasks

While the principal task was to conduct a global literature search in order to identify attributes as reflected in the specified task above, it was observed that it might be beneficial to consider approaches for presenting relative hazard or risk associated with green building elements and features, especially if such information was not identified as part of the literature search. As such, the project team proposed to pursue several additional activities as outlined below:

- Develop a list of fire and life safety attributes of concern against which to compare or assess the relative safety performance of green building elements (e.g., toxicity, flame spread, smoke spread, impact of fire-fighting operations, etc.).
- Review LEED and other such systems to understand implication of fire and safety performance objectives on green ratings or compliance requirements.
- Develop set of green and 'safety' performance objectives (metrics if possible). This is needed to assess combined performance.

- In addition to the literature search, survey various building regulatory agencies and fire research institutions around the world to obtain their experiences with fire and green buildings.
- Compile from the information search and survey existing research studies related to incorporating building safety, life safety and fire safety as an explicit element in green building indices, identifying gaps and specific needed research areas identified in those reports.
- Develop a first-order risk matrix/risk ranking approach that couples green building elements, products, systems and features with identified fire or life safety impacts of concern, benchmarking initially against 'conventional' construction and experience, using a relative 'greater than,' 'equivalent to,' or 'less than' approach, based in the first instance on expert judgment.
- Present a 'first-order' risk and performance assessment tool (matrix, perhaps) which incorporates green and safety performance indicators, and allows for future systematic assessment and documentation of green building design elements that may increase fire safety hazards.

1.3 Research Direction and Observations

Research was conducted as outlined above, beginning with a literature search, followed up by surveys, and including steps towards development of a hazard or risk ranking matrix. Research included green building site issues in addition to green building elements, since fire safety can be impacted by fire fighter access the site and building. Since few incidents of fires in documented green buildings were identified, there is currently insufficient data for quantified risk assessment. Also, since only limited data were found on actual fire performance tests of green building elements, and no data were found on comparative hazard analysis, a qualitative risk/hazard ranking approach was ultimately pursued. As presented, the relative rankings are based on expert judgment, and a much more formal process is recommended for obtaining stakeholder input on the hazard/risk factors and establishing rankings should future work be undertaken on this project. In addition, since this effort did not consider the green or sustainability benefits of fire protection measures, it is recommended that this aspect be included in future research associated with this FPRF project.

Chapter 2
Information Search

Various searches and inquiries have been undertaken by the project team with the aim to identify fire incident reports, news reports, fire investigation reports, and research reports related to fires in green buildings and fires involving green building elements. These include web-based searches using generic search platforms (e.g., Google), targeted searches supported by WPI library staff (e.g., LEXIS/NEXIS), and searches of research and academic institution holdings (e.g., NIST, NRCC, BRANZ, WPI, etc.). The searches by the project team were supplemented by searches conducted by NFPA Research staff, inquires sent to the Technical Panel for this project, and inquires sent to the FPRF Property Insurance Research Group which sponsored this project. In addition, targeted inquiries were made via the IRCC (a group of 14 building regulatory agencies in 12 countries, www.irccbuildingregulations.org) and the Fire FORUM (an international group of fire research laboratory directors, http://www.fireforum.org/). Representative findings are provided below, with additional information in the appendices.

2.1 Representative Fire Incidents

In order to identify as many fire incidents involving green buildings and building elements as practicable, the project team reached out to several entities in the USA and internationally, including building regulatory agencies, fire service entities, insurance companies and research entities. The first stage involved web-based searches and requests via NFPA for fire incident data base searches. As a result of these searches a few dozen incidents were identified. A representative selection of incidents is presented in Table 2.1. While relatively small in number when compared to all fires, these incidents reflect a diverse set of fire and green building/element related issues, and helped form the basis of attributes identified and considered in Table 3.1 (Chap. 3) and Table 4.1 (Chap. 4).

B. Meacham et al., *Fire Safety Challenges of Green Buildings*, SpringerBriefs in Fire, DOI: 10.1007/978-1-4614-8142-3_2, © Fire Protection Research Foundation 2012

Table 2.1 Representative fire incidents

Commercial photovoltaic panel fire	
383 kW roof PV system fire, Target Store, Bakersfield, CA, April 2009	http://nfpa.typepad.com/files/target-fire-report-09apr29.pdf (last accessed 10/21/12)
PV roof fire, France warehouse, January 2010	http://www.aria.developpement-durable.gouv.fr/ressources/fd_37736_valdereuil_jfm_en.pdf (last accessed 10/21/12)
Roof PV system in Goch, Germany, April 2012	http://www.feuerwehr-goch.de/index.php?id=22&tx_ttnews%5Btt_news%5D=596&cHash=982afcd5c431b7299f67de4af397cc43 (last accessed 10/21/12)
1,208 kW roof PV system, Mt. Holly, NC, April 2011	http://www.solarabcs.org/about/publications/meeting_presentations_minutes/2011/12/pdfs/Duke-Webinar-Dec2011.pdf (last accessed 10/21/12)
PV roof fire, Trenton, NJ, March 2012	http://blog.nj.com/centraljersey_impact/print.html?entry=/2012/03/trenton_firefighters_battle_ro.html (last accessed 10/21/12)
	http://www.nj.com/mercer/index.ssf/2012/03/solar_panels_source_of_fire_at.html (last accessed 10/21/12)
Residential photovoltaic panel fire	
PV Fire: Experience and Studies, UL, 2009	http://www.solarabcs.org/about/publications/meeting_presentations_minutes/2011/02/pdfs/Arc-PV_Fire_sm.pdf (last accessed 10/21/12)
PV fires, FPRF report, 2010	http://www.nfpa.org/assets/files/pdf/research/fftacticssolarpower.pdf (last accessed 10/21/12)
PV fire, San Diego, CA, April 2010	http://www.nctimes.com/article_8a32fb03-9e3f-58ca-b860-9c7fe1e28c7e.html (last accessed 10/21/12)
PV fire, Stittingbourne, UK, March 2012	http://www.kentonline.co.uk/kentonline/news/2012/march/30/solar_panels.aspx (last accessed 10/21/12)
Battery storage and UPS fire	
Battery fire, Data Center, Taiwan, February 2009	http://indico.cern.ch/getFile.py/access?sessionId=8&resId=1&materialId=0&confId=45473 (last accessed 10/21/12)
Residential spray foam insulation fire	
Foam insulation home fire, North Falmouth, MA, May 2008	http://www.capecodonline.com/apps/pbcs.dll/article?AID=/20080520/NEWS/805200318/-1/rss01 (last accessed 10/21/12)
	http://www.greenbuildingadvisor.com/blogs/dept/green-building-news/three-massachusetts-home-fires-linked-spray-foam-installation (last accessed 10/21/12)
Foam insulation, Woods Hole, MA, February 2011	http://www.capecodonline.com/apps/pbcs.dll/article?AID=/20110211/NEWS/102110323 (last accessed 10/21/12)

(continued)

Table 2.1 (continued)

Commercial photovoltaic panel fire	
Foam insulation fire, Quebec, May 2010	http://www.greenbuildingadvisor.com/blogs/dept/green-building-news/nze-project-tragic-fire-and-will-rebuild (last accessed 10/21/12)
Residential foil insulation, fire/shock hazards	
Home insulation program (Australia)	http://www.climatechange.gov.au/government/initiatives/hisp/key-statistics.aspx (last accessed 10/21/12)
	http://www.productsafety.gov.au/content/index.phtml/itemId/974027/ (last accessed 10/21/12)
	http://www.wsws.org/articles/2010/feb2010/insu-f22.shtml (last accessed 10/21/12)
	http://www.theaustralian.com.au/news/garretts-roofing-fire-admission/story-e6frg6n6-1225829880090 (last accessed 10/21/12)
Exterior finish and insulation systems fire	
The Monte Carlo Exterior Façade Fire (2008)	http://usatoday30.usatoday.com/news/nation/2008-01-25-vegas-fire_N.htm (last accessed 10/21/12)
	http://magazine.sfpe.org/fire-investigation/monte-carlo-exterior-facade-fire (last accessed 10/21/12)
Sandwich panels/structural integrated panel (SIP) with combustible foam insulation or coating	
Borgata Casino, Atlantic City, NJ, Façade Fire (2007)	http://www.fireengineering.com/articles/2010/05/modern-building-materials-are-factors-in-atlantic-city-fires.html (last accessed 10/21/12)
Apartment Façade Fire, Busan, Korea	http://koreabridge.net/post/haeundae-highrise-fire-busan-marine-city-burns (last accessed 10/21/12)
	http://view.koreaherald.com/kh/view.php?ud=20101001000621&cpv=0 (last accessed 10/21/12)
Apartment Façade and Scaffold Fire, Shaghai, China	http://www.boston.com/bigpicture/2010/11/shanghai_apartment_fire.html (last accessed 10/21/12)
	http://www.bbc.co.uk/news/world-asia-pacific-11760467 (last accessed 10/21/12)
High-Rise Façade Fires, UAE	http://gulfnews.com/news/gulf/uae/emergencies/fire-breaks-out-at-sharjah-tower-1.1014750 (last accessed 10/21/12)
	http://www.emirates247.com/news/emirates/dh50-000-fine-for-fire-safety-violation-in-high-rises-2012-05-07-1.457534 (last accessed 10/21/12)
	http://article.wn.com/view/2012/05/02/Municipality_moves_to_ban_flammable_tiles/ (last accessed 10/21/12)
	http://article.wn.com/view/2012/05/02/Tower_cladding_in_UAE_fuels_fire/ (last accessed 10/21/12)
	http://article.wn.com/view/2012/05/01/Experts_shed_light_on_how_fires_spread_in_towers/ (last accessed 10/21/12)
Façade Fire, Beijing, China	http://www.nytimes.com/2009/02/10/world/asia/10beijing.html?_r=1 (last accessed 10/21/12)

2.2 Selected Resources Related to Fire and Green Building Concerns

In addition to identifying fire incidents, the project team was also interested in identifying fire-related concerns with green buildings and building elements. The starting point for this search was also web-based searches, considering general media, trade publications, peer review articles, and research reports. Much like the incident data, the number of publications/resources identified is somewhat low. This is in part due to challenges associated with web searches, limited responses to inquiries (see survey section), and general lack of efforts on fire and green building issues *defined as such*. This latter point is important, as some of the research identified by the project team has been attributed by the team as being related to green building issues, but might not have been by entities which conducted research that is cited (e.g., UL investigation into LEL and structural stability concerns, which was more closely identified as a fire fighter safety issue). This type of confounding representation likely means more research is available, but requires more effort to identify. Nonetheless, Table 2.2 contains a representation of the types of articles, reports and studies related to fire and green building concerns.

While details can be found via the links provided, selected incidents, test programs and mitigation approaches are summarized in Appendix B. In addition, the following resources provide significant discussion relative to the project focus, and are highly recommended as key sources of information on the topic of green buildings and fire:

- The BRANZ study, *Building Sustainability and Fire-Safety Design Interactions*, http://www.branz.co.nz/cms_show_download.php?id=716733515027fe462618 8881f674635d51e3cfb0 (last accessed 10/21/12)
- The BRE study, *Impact of Fire on the Environment and Building Sustainability*, http://www.communities.gov.uk/documents/planningandbuilding/pdf/1795639. pdf (last accessed 10/21/12)
- The NASFM Green Buildings and Fire Safety Project (report and web links), http://www.firemarshals.org/programs/greenbuildingsandfiresafetyprojects.html (last accessed 10/21/12)

In addition, it is worth noting that two of the above studies, those by BRE and BRANZ, also address the contribution of fire protection measures to sustainability. While this effort did not consider this topic, it is an area that should be considered in the overall assessment of building fire safety and sustainability. In this regard, studies undertaken by FMGlobal contribute significantly in this area as well:

- *The Influence of Risk Factors on Sustainable Development*—http://www.fmglob al.com/assets/pdf/P09104a.pdf (last accessed 10/21/12)
- *Environmental Impact of Fire Sprinklers*—http://www.fmglobal.com/assets/pdf/ P10062.pdf (last accessed on 10/21/12—registration may be necessary)

Table 2.2 Fire safety concerns in green buildings: Selected resources

Overall concerns	
BRANZ—Building sustainability and fire-safety design interactions (2012)	http://www.branz.co.nz/cms_show_download.php?id=71673351 5027fe462618 8881f674 635d51e3cfb0 (last accessed 10/21/12)
BRE—Impact of fire on the environment and building sustainability (2010)	http://www.communities.gov.uk/documents/planningandbuilding/pdf/1795639.pdf (last accessed 10/21/12)
Green fire initiatives—Links to related studies, National Association of State Fire Marshals (2010)	http://www.firemarshals.org/greenbuilding/greenfireinitiatives.html#greenroofs (last accessed 10/21/12)
Bridging the gap: Fire safety and green buildings, NASFM 2010	http://firemarshals.org/greenbuilding/bridgingthegap.html (last accessed 10/21/12)
Fire safety green buildings, an IQP project by Joyce, Miller, Wamakima (WPI 2008)	http://www.wpi.edu/Pubs/E-project/Available/E-project-121908-111921/unrestricted/ Final_IQP_Report.pdf (last accessed 10/21/12)
Photovoltaic/energy systems	
Fire operations for photovoltaic emergencies, CAL fire-office State fire Marshal, 2010	http://osfm.fire.ca.gov/fromthechief/pdf/sfmreportnov10.pdf (last accessed 10/21/12)
Fire fighter safety and emergency response for solar power systems	http://www.nfpa.org/assets/files/pdf/research/fftacticssolarpower.pdf (last accessed 10/21/12)
Firefighter safety and photovoltaic installations research project, UL 2011	http://www.ul.com/global/documents/offerings/industries/buildingmaterials/fireservice /PV-FF_SafetyFinalReport.pdf (last accessed 10/21/12)
The ground-fault protection blind spot: A safety concern for larger photovoltaic systems	http://www.solarabcs.org/about/publications/reports/blindspot/pdfs/BlindSpot.pdf (last accessed 10/21/12)
Lightweight wood structures	
Lightweight structure fire, NFPA	http://www.nfpa.org/publicJournalDetail.asp?categoryID=1857&itemID=43878&src= NFPAJournal&cookie%5Ftest=1 (last accessed 10/21/12)
Structural collapse under fire conditions, Toomey 2008	http://www.fireengineering.com/articles/print/volume-161/issue-5/departments/training-notebook/structural-collapse-under-fire-conditions.html (last accessed 10/21/12)
Improving fire safety by understanding the fire performance of engineered floor systems and providing the fire service with information for tactical decision making, UL 2012	http://www.ul.com/global/documents/offerings/industries/buildingmaterials/fireservice/ basementfires/2009 NIST ARRA Compilation Report.pdf (last accessed 10/21/12)

(continued)

Table 2.2 (continued)

Overall concerns	
Structural collapse: the hidden dangers of residential fires (Dalton)	http://www.fireengineering.com/articles/print/volume-162/issue-10/features/structural-collapse.html (last accessed 10/21/12)
Architectural	
Fire safety concern on well-sealed green buildings with low OTTVs (Chow 2010)	http://docs.lib.purdue.edu/cgi/viewcontent.cgi?article=1000&context=ihpbc (last accessed 10/21/12)
A short note on fire safety for new architectural features (Chow 2004)	http://www.bse.polyu.edu.hk/researchCentre/Fire_Engineering/summary_of_output/journal/IJAS/V5/p.1-4.pdf (last accessed 10/21/12)
Architectural	
Performance of double-skin façade	http://www.bse.polyu.edu.hk/researchCentre/Fire_Engineering/summary_of_output/journal/IJEPBFC/V6/p.155-167.pdf (last accessed 10/21/12)
Window reflecting melting vinyl siding	http://www.greenbuildingadvisor.com/blogs/dept/musings/window-reflections-can-melt-vinyl-siding (last accessed 10/21/12)
Fire hazards of foam insulation	
Exterior walls, foam insulating materials, and property risk considerations (2007)	http://www.risklogic.com/articles/may2007.html (last accessed 10/21/12)
Panelized Construction problems	http://www.njeifs.com/lawyer-attorney-1513025.html (last accessed 10/21/12)
Toxicity of flame retardants in foam insulation and other products	
Brominated flame retardants and health concerns	http://www.ncbi.nlm.nih.gov/pmc/articles/PMC1241790/ (last accessed 10/21/12)
PBDE flame retardants/potential adverse health effects	http://www.actabiomedica.it/data/2008/3_2008/costa.pdf (last accessed 10/21/12)
Toxicity of flame retardants and impact on fire fighters	http://www.nist.gov/el/fire_research/upload/4-Purser.pdf (last accessed 10/21/12)
Industrialized roof farming	
Rooftops take urban farming to the skies	http://today.msnbc.msn.com/id/32643514/ns/today-green/t/rooftops-take-urban-farming-skies (last accessed 10/21/12)
Wind farm	
Dark side of green wind turbine accidents	http://eastcountymagazine.org/print/9238 (last accessed 10/21/12)

It is recommended that future research into the interactions of fire protection and building sustainability include consideration of the potential benefits of fire protection to sustainability as well as the potential detriments of green construction to fire and life safety.

2.3 International Survey and Responses

In addition to the web-based searches, a number of targeted inquiries were sent out, including requests for information sent to the NFPA Research Division, the FPRF panel members and their organizations, member countries of the Inter-jurisdictional Regulatory Collaboration Committee (IRCC) and associated organizations in their countries (i.e., fire service, research or insurance entities to which they forwarded requests and/or provided contact information), and members of the International FORUM of Fire Research Directors (the FORUM). Response was reasonable (e.g., NFPA, several FPRF panel members and 8 of 14 IRCC members responded); however, data were limited, since in all cases it was reported that data specific to fire in green buildings is not being tracked, as this criterion is not including in existing fire incident reporting systems (including NFIRS in the USA). That challenge aside, some data on fires involving green building elements was provided, such as by the New South Wales Fire Brigade (Australia) as reflected in Table 2.3 (see Appendix C for complete survey responses).

2.4 Review of Representative Green Building Rating Schemes for Fire Considerations

The problem statement for this project noted the proliferation of rating schemes for green buildings and the development of green building and construction codes which promote the use of green materials and systems, but which perhaps do not consider fire safety concerns, and that "a systematic method needs to be developed for implementation in the certification process that integrates the consideration of fire as well as other hazard risk factors as part of design performance metrics." In order to make progress on this, not only is it required to understand what constitutes green buildings and elements, and what fire hazards or risks they might pose, it is important to understand in which areas the existing rating schemes and codes might be imposing unintended fire safety consequences.

The information search revealed that globally there more than two dozen green building rating schemes available (e.g., see http://www.gsa.gov/graphics/ogp/sustainable_bldg_rating_systems.pdf, last accessed on 10/29/12). In addition, several systems have multiple schemes by building use, such as retail, school, residential, office, etc., and some include separate schemes for new and existing buildings. Likewise, there are a number of green building codes world-wide, including the

Table 2.3 Survey responses from New South Wales fire brigade

Country/entity	Fire Incident experience/tracking in green buildings	Fire Incident experience/tracking Involving green building elements	Risk-based assessment of green building elements
Australia			
New South Wales Fire Brigade	The structures that subscribe to the National Built Environment Rating System (NABERS) are usually commercial or government buildings. In most cases they are relatively new and range from modern high rise premises in the city (e.g., No. 1 Bligh St.) to restored and renovated federation style buildings (e.g., 39 Hunter St.) The building codes also provide for prescribed or engineered fire safety solutions. There are no specific AIRS codes for "Green" buildings therefore it is very difficult to determine if there have been any fires or dominant fire causes in these buildings	*Ceiling Insulation*: FIRU have experienced major concerns with this issue particularly in residential, nursing homes and aged care facilities. Cellulose fibre insulation in close proximity to down lights and insulation batts including non-compliance with electrical wiring rules have been the dominant concerns. AIRS analysis for insulation fires 29/02/2008 to 22/06/2011. Data provided by SIS: The data includes 102 incidents that occurred in metropolitan, regional and country areas. Of these incidents 75 were directly related to down lights and their associated transformers in close proximity to ceiling insulation. Some of the fires occurred in substantial property damage. *Insulated Sandwich Panels*: No specific AIRS codes for Insulated Sandwich Panels. FIRU have reports of residential structures constructed of insulated sandwich panels in locations ranging from Broken Hill to Thredbo. *Laminated Timber I-Beams*: No specific AIRS codes for Laminated Timber I-Beams. A combination of open plan living and modern furnishings (e.g. polyurethane foam settees, etc.) can create fuel packages that will reach temperatures of 1,000–1,200 °C that can rapidly weaken structural elements. *Photovoltaic Solar Installations*: No specific AIRS codes for PV solar installations. It is estimated that NSW has about 150,000 PV solar installations. Predominant causes include faulty components and incorrect installation. (2010 to August 2012 = nine reported incidents) FIRU research indicates major problems with PV solar installations. No remote solar isolation switching, no DC rest device for Firefighters and no 24/7 availability of qualified PV solar electricians. This has been reflected in reports from around Australia, Germany and the US. From FIRU research it appears that major fire services throughout Australasia, United Kingdom, Germany and the United States have experienced concerns with "Green" building elements. When attending fire calls operational Firefighters are unable to readily determine if it is a "Green" building or one with "Green" building elements	FRNSW has a Risk Based Approach for all incidents, including all types of structures. This is supported by a broad range of Standard Operational Guidelines (SOGs), Safety Bulletins and Operations Bulletins. No specific effort related to green buildings

IgCC, the Code for Sustainable Homes in England (http://www.planningportal. gov.uk/uploads/code_for_sust_homes.pdf, last accessed on 10/29/12) and others.

Only a small subset of the available green rating systems was able to be reviewed within the bounds and scope of this project. The sample of green rating schemes selected for this project was determined based on freely available information. The sample ultimately included LEED (residential and retail), BREEAM (new buildings), GREEN MARK (residential and nonresidential), and the IgCC. More discussion on the review and findings relative to these schemes can be found in Chap. 5 and Appendix G.

Review of this sample of green building rating schemes and the IgCC indicated that fire safety objectives are not explicitly considered in these systems. This is not unexpected, however, since the focus is principally on resource efficiency (e.g., energy, water, materials) and not on safety. In the case of BREEAM, a study by BRE (BRE BD2709 2010, p 45) notes that "fire safety and fire protection are not included in most BREEAM schemes since most BREEAM schemes assess new buildings and the BREEAM assessment takes for granted that the building will satisfy the Building Regulations; the BREEAM assessment relates to *additional* sustainability features."

Although no specific references regarding fire safety objectives were identified in LEED and GREEN MARK documentation, it is hypothesized that similar rationale applies as with BREEAM. With a voluntary system, which aims to encourage sustainable practices, it is anticipated that basic building code requirements, including fire safety, are met via code compliance. This is also the case with the IgCC, which is intended to work along with the International Building Code (IBC) and relevant codes and standards. If one then assumed that the risks or hazards associated with green building elements and features are addressed adequately by building codes and standards, one could assume that no additional risk or hazards exist. However, it can be that current fire tests, which have been determined as adequate for conventional construction, may not yet be fully vetted for innovative and green construction with respect to performance in use (e.g., LEL). Further study is recommended in this area.

Although none of the green building rating schemes that were reviewed during this project included fire safety objectives, it was found that the scheme of the German Sustainable Building Council (DGNB) includes criteria for fire prevention (http://www.dgnb-system.de/dgnb-system/en/system/criteria/, accessed last on 10/29/12). Although detail on the weights of fire prevention attributes relative to the green attributes was not able to be verified it is understood that some credit is given for fire protection features such as smoke extract, automatic sprinklers, and structural fire protection.

Likewise, it was determined that BREEAM-in-USE (http://www.breeam.org/ page.jsp?id=373, last accessed on 10/29/12), a recent BRE scheme to help building managers reduce the running costs and improve the environmental performance of existing buildings, incorporates fire risk reduction attributes (BRE 2010). The fire risk reduction attributes related to such issues as whether a fire risk assessment has been conducted, are emergency plans in place, and so forth. No indication of consideration of fire protection systems was identified.

Chapter 3
Green Building/Site Elements and Attributes

Given the focus of the study, a necessary step was to identify green building features to consider. The following table reflects green building elements and attributes that were selected for this effort based on the literature review and survey responses (see Appendices A, B and C). While this list is extensive, it may not be exhaustive, and it is recommended that the list be updated by future studies as knowledge of other green materials, features, elements and attributes is identified (Table 3.1).

Table 3.1 Green building/site elements and attributes

Structural materials and systems	Interior materials and finishes	Alternative energy systems
• Lightweight engineered lumber	• FRP walls/finishes	• PV roof panels
• Lightweight concrete	• Bio-polymer wall/finishes	• Oil-filled PV panels
• Fiber reinforced polymer (FRP) elements	• Bamboo walls/finishes	• Wind turbines
• Plastic lumber	• Wood panel walls/finishes	• Hydrogen fuel cells
• Bio-polymer lumber	• Bio-filtration walls	• Battery storage systems
• Bamboo	• Glass walls	• Cogeneration systems
• Phase-change materials	• FRP flooring	• Wood pellet systems
• Nano materials	• Bio-polymer flooring	• Electric vehicle charging station
• Extended solar roof panels	• Bamboo flooring	• Tankless water heaters
Exterior materials and systems	*Interior space attributes*	*Site issues*
• Structural integrated panel (SIP)	• Tighter construction	• Permeable concrete systems
• Exterior insulation & finish (EFIS)	• Higher insulation values	• Permeable asphalt paving
• Rigid foam insulation	• More enclosed spaces	• Use of pavers
• Spray-applied foam insulation	• More open space (horizontal)	• Extent (area) of lawn

(continued)

B. Meacham et al., *Fire Safety Challenges of Green Buildings*, SpringerBriefs in Fire, DOI: 10.1007/978-1-4614-8142-3_3, © Fire Protection Research Foundation 2012

Table 3.1 (continued)

• Foil insulation systems	• More open space (vertical)	• Water catchment/features
• High-performance glazing	• Interior vegetation	• Vegetation for shading
• Low-emissivity & reflective coating	• Skylights	• Building orientation
• Double-skin façade	• Solar tubes	• Increased building density
• Bamboo, other cellulosic	• Increased acoustic insulation	• Localized energy production
• Bio-polymers, FRPs	*Building systems and issues*	• Localized water treatment
• Vegetative roof systems	• Natural ventilation	• Localized waste treatment
• PVC rainwater catchment	• High volume low speed fans	• Reduced water supply
• Exterior cable/cable trays	• Refrigerant materials	• Hydrogen infrastructure
Façade Attributes	• Grey-water for suppression	• Community charging stations
• Area of glazing	• Rain-water for suppression	
• Area of combustible material	• On-site water treatment	
• Awnings	• On-site waste treatment	
• Exterior vegetative covering	• On-site cogeneration	
	• High reliance on natural lighting	
	• PV exit lighting	
	• Reduced water suppression systems	

Chapter 4
Attributes of Green Building or Site Which Could Impact Fire, Life Safety, Building or Fire Service Performance

In order to assess relative increases in fire hazard or risk or decreases in safety or performance of green building elements or attributes as compared with conventional construction, a list of risk, hazard or performance attributes of concern was required. The list in Table 4.1 was compiled from a combination of fire and life safety performance objectives typically addressed by building and fire codes and from issues identified during the literature review. This list reflects a focus on occupant and emergency responder safety issues and building performance issues. The list does not explicitly consider building contents protection, business continuity, or related market issues, which may also be of concern. While the list of attributes might be expanded or refined in the future, it provides a reasonable starting point and basis for comparative analysis.

The lists in Tables 3.1 and 4.1 were used in the development of matrices which could potentially be used as a checklist to help review a building plan, a building and/or a building site for potential risks or hazards, as well as a mechanism to reflect relative risk level associated with the green building element. These matrices/tools are discussed in Sects. 5.2 and 5.3 respectively and detailed in Appendices D and E. The information in Tables 3.1 and 4.1 also formed the basis of the relative hazard and mitigation matrix discussed in Sect. 5.1.

B. Meacham et al., *Fire Safety Challenges of Green Buildings*, SpringerBriefs in Fire, DOI: 10.1007/978-1-4614-8142-3_4, © Fire Protection Research Foundation 2012

Table 4.1 Hazard, risk and performance attributes

Poses potential ignition hazard
Poses potential shock hazard
Poses potential explosion hazard
Poses potential toxicity hazard
Readily ignitable
Burns readily once ignited
Contributes more fuel/increased heat release rate (HRR)
Material affects burning characteristics
Fast(er) fire growth rate
Significant smoke production/hazard
Potential for shorter time to failure
Failure affects burning characteristics
Failure presents smoke spread concern
Failure presents flame spread concern
Material presents flame spread concern
May impact smoke/heat venting
May impact occupant evacuation
May impact fire-fighter (FF) water availability
May impact suppression effectiveness
May impact fire apparatus access
May impact fire-fighter (FF) access and operations
May impact containment of runoff

Chapter 5
Hazard/Risk Assessment and Ranking

There are various approaches to fire hazard and risk assessment and ranking, from qualitative to quantitative. Detailed discussion of these can be found in many sources, including the *SFPE Handbook of Fire Protection Engineering* (NFPA 2008), *SFPE Engineering Guide on Fire Risk Assessment* (SFPE 2006), NFPA 551, *Guide for the Evaluation of Fire Risk Assessments* (NFPA 2012), textbooks (e.g., Ramachandran and Charters 2011), and in the literature (e.g., Meacham 2004; Meacham et al. 2012).

Given the scope, data, time and resources for this effort, the approach taken was qualitative, both for hazard/risk identification and level of severity (concern, importance). A principal driver for qualitative risk assessment is the lack of data on fires in green buildings and elements. Based on surveys and searches, needed data are not being collected systematically, and the number of incidents identified is small in number. Likewise, for the hazard assessment, while there are detailed assessments of specific building systems and elements, such as UL and NRC Canada research on lightweight engineered lumber (LEL) performance under fire conditions (e.g., see http://www.nrc-cnrc.gc.ca/eng/ci/v16n2/1.html, http://www.ul.com/global/eng/pages/ offerings/industries/buildingmaterials/fire/fireservice/lightweight/), the data and the applications are limited with respect to the population of buildings which might be impacted. While generalizations can be extracted, such as certain floor systems using unprotected LEL joists between the basement level and first floor level of a single family dwelling failed more quickly than typical sawn lumber wood construction (i.e. 2 × 4 construction) for the scenarios and fires tested, it is difficult to quantify the actual difference in hazard or risk in comparison to the conventional wood construction without significantly more analysis. However, if such detailed analysis is undertaken, that will be a large step towards quantitative hazard assessment. Such an activity is recommended for the future.

The above discussion relative to fire performance of LEL floor systems as compared with 'conventional' construction is indicative of another challenge: assessing the relative increase in fire hazard or risk, or decrease in safety performance, associated with green building elements, even if they comply with current code requirements and associated test requirements. On the one hand, it is easy to say that any

B. Meacham et al., *Fire Safety Challenges of Green Buildings*, SpringerBriefs in Fire, DOI: 10.1007/978-1-4614-8142-3_5, © Fire Protection Research Foundation 2012

product, whether it green or conventional, achieves the same level of required performance as dictated by current regulatory requirements. However, as is being indicated by the LEL floor system research, while individual elements and products
might achieve minimum test and code compliance, there can still be a difference
in safety performance in relation to the product in use as part of a larger building system. This is not to say that forcing all green products to comply with current building code requirements and fire test requirements is in any way bad: quite
the contrary, it is a very good first step and is a solid mitigation strategy. However,
because of the intrinsic properties of some green elements, such as less material
(e.g., LEL, or lighter weight high-strength concrete), or modified properties relative
to energy performance which could increase fire or safety hazards beyond typical
conventional system installations (e.g., more insulation, if combustible, adds the
potential for additional fire load, even if a fire barrier is required; therefore, the magnitude of fire hazard would be larger than conventional systems if the thermal barrier
fails), there are identifiable areas of concern that are currently not quantified. In-use
configuration and fire scenarios are clear areas for further research and development.

Even when taking a qualitative approach to hazard/risk ranking, as done in this
report, there are several ways in which the information can be presented and different depths at which the hazard or risk factors can be addressed relative to the green
building issues identified. In the following sections three representations are provided: a detailed matrix with green element and attributes and potential hazard/risk
factors; a tabular approach with fewer attributes but more discussion; and a qualitative pictorial indication of potential fire hazards within green rating schemes. Each
of these approaches reflects a different area of focus. The detailed matrix is provided as the basis for a potential hazard/risk ranking tool. The user might see an
input screen, where she indicates specific green elements or site features associated with a project, and the outcome would identify specific areas of concern and
relative level of increased risk or hazard or decreased safety performance. This
would then require mitigation options to be explored and implemented as desired.
The tabular format presents the risk/hazard concerns in narrative form, provides
a subjective ranking of relative importance level, and presents general mitigation
strategies (e.g., use approved products, provide thermal barrier, provide sprinkler
protection, etc.). This could be used for policy-level decisions or to guide selection
of mitigation strategies. The pictorial approach might serve as the basis for some
type of quick overview guide or energy and fire performance ranking for a building. All of these approaches are currently subjective and based on the views of the
project team, and should be taken as illustrative only.

Building on the issue of limited input for the rankings as presented in this report,
another complication at this stage is the lack of broad stakeholder participation to
reach agreed upon risk/hazard attributes and relative risk/hazard rankings that would
be required for implementable tools. The need for broad stakeholder participation in
achieving consensus is widely understood, since whether or not increased hazard (or
risk) is present, and if so, to whom or what, will depend on the stakeholder position (e.g., see Meacham 2004; AS/NZS4630 2004; ISO 31000 2009; Meacham
et al. 2008; Watts 2008). In the above discussion, for example, manufacturers of LEL

systems would likely have a much different perspective than the fire service. To get a balanced and agreed upon set of risk/hazard rankings, broad stakeholder participation, developed within a structured risk/hazard ranking exercise/environment, would be needed. Various approaches for conducting such as process are described in the literature (e.g., NAP 1996; Meacham 2000; Watts 2008; Ramachandran and Charters 2011). This is a recommended are for future development. Since such a broad stakeholder exercise was not undertaken as part of this effort, the hazard/risk/performance rankings as expressed within this report should be taken as illustrative only.

5.1 Detailed Matrices of Green Building Elements/Features and Hazard/Risk Factors

As noted in Chap. 4, a list of potential fire hazards/risks associated with the green building elements and attributes identified in Chap. 3 was developed. These can be combined into matrices which can be used as an assessment tool and as a hazard or risk presentation tool. An example of the combined attribute—fire hazard matrix for *Exterior Materials and Systems* is shown in Fig. 5.1. A complete set of matrices by building/site area can be found in Appendix D.

It is suggested that a blank matrix, as illustrated in Fig. 5.1, could serve as a checklist for engineers, designers, insurers, authorities or others when reviewing site plans, building designs, renovation designs or buildings to guide inspection of green attributes which could result in a fire hazard or building fire performance concern. It does not, however, give any indication of the relative magnitude of the increased hazard or decreased performance. However, the matrix has the potential to be developed into a hazard/performance ranking tool, where the user identifies the green element or attribute in the design or building, and the tool provides an indication as to whether any increased fire risk or decrease performance might be expected. A potential approach to presenting relative risk/hazard information that could be generated by such a tool is illustrated in Fig. 5.2.

In this figure, a relative level of fire risk or hazard is illustrated, wherein the blank boxes reflect low (or not applicable) risk or hazard, yellow boxes reflect moderate risk or hazard, and red reflects high risk or hazard, *in comparison to conventional element, systems and features*, and where no mitigation measure has been implemented. (As discussed earlier in the report, some level of mitigation is provided by compliance with existing building regulations and test standards; however, more research is needed to explore the degree of mitigation as compared with conventional construction in the context of the design and fire scenario). Note that as discussed before, the assigned levels are all based on limited expert judgment at this time and should be taken as illustrative only. It is conceivable that a tool like this can be developed for engineers, designers, insurers, authorities or others when reviewing site plans, building designs, renovation designs or buildings to assist in hazard or risk analysis, and for developing mitigation strategies. Depending on what mechanism is used to create the underlying estimation of

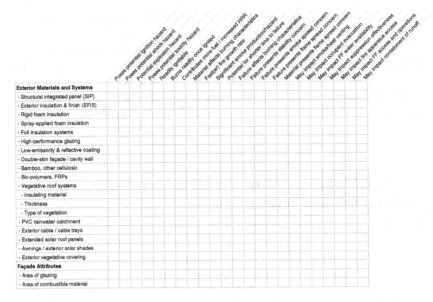

Fig. 5.1 Matrix of green attributes and potential fire hazards

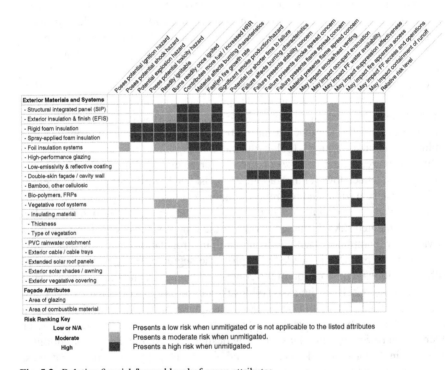

Fig. 5.2 Relative fire risk/hazard level of green attributes

relative risk,[1] this format lends itself to development into a spreadsheet tool. A complete set of exemplar matrices by building/site area can be found in Appendix E.

5.2 Tabular Representation of Potential Fire Hazards with Green Building Elements

While the presentation of information as reflect in the matrices in Sect. 5.1 have the benefit of details associated with green elements and features, it has the shortcomings of lack of description of how the hazard or risk manifests and what potential mitigation strategies might be. In order to combine all of this information, we chose to develop a tabular representation of the green element, fire hazard (or risk), level of concern, and potential mitigation strategies. This is illustrated in Fig. 5.3 for the same *Exterior Materials and Systems* building areas used in the previous section. Although this is subjective at this point, the techniques overviewed before can be applied to result in a more quantitatively-based representation. A complete table for all building/site areas can be found in Appendix F.

5.3 Fire Hazards Associated with Green Rating Schemes and Codes

This aspect of the project involved review of various green building rating schemes, as well as the International Green Construction Code (IgCC), to determine (a) if and how fire and life safety objectives are included, and (b) if and how green attributes considered by the scheme may affect fire and life safety performance. Globally there more than two dozen green building rating schemes available for use (e.g., see http://www.gsa.gov/graphics/ogp/sustainable_bldg_rating_systems.pdf, last accessed on 10/29/12). In addition, several systems have multiple schemes by building use, such as retail, school, residential, office, etc., and some include separate schemes for new and existing buildings. Unfortunately, only a small subset of these systems was able to be reviewed within the bounds and scope of this project. Likewise, there are a number of green building codes world-wide, including the International Green Construction Code (IgCC), the Code for Sustainable Homes in England (http://www.planningportal.gov.uk/uploads/code_for_sust_homes.pdf, last accessed on 10/29/12) and others.

The sample of green building rating schemes selected for this project was determined based on freely available information. The sample ultimately included LEED

---

[1] For example, a semi-qualitative approach could be used where 'low,' 'moderate' and 'high' risk are assigned numerical values, and the 'relative risk level' is estimated as a weighted function of the importance or influence of hazard impact on the various 'green' elements.

Material / System / Attribute	Hazard	Concern Level	Potential Mitigation Stratgies
Exterior Materials and Systems			
- Structural integrated panel (SIP)	If fail, insulation can contribute to flame spread, smoke production and fuel load.	High	Approved / listed materials. Assure proper sealing of panels. Take care during installation, including retrofits, relative to potential sources of ignition.
- Exterior insulation & finish (EFIS)	If fail, insulation can contribute to flame spread, smoke production and fuel load.	High	Approved / listed materials. Assure proper sealing of panels. Take care during installation, including retrofits, relative to potential sources of ignition.
- Rigid foam insulation	Can contribute to flame spread, smoke and toxic product development and fuel load.	High	Fire resistive barrier (e.g., fire rated gypsum). Approved / listed materials. Flame retardants. Sprinklers.
- Spray-applied foam insulation	Can contribute to flame spread, smoke and toxic product development and fuel load.	High	Fire resistive barrier (e.g., fire rated gypsum). Approved / listed materials. Flame retardants. Sprinklers.
- Foil insulation systems	Can contribute to shock hazard for installers. Can contribute to flame spread and fuel load.	High	Fire resistive barrier (e.g., fire rated gypsum). Approved / listed materials. Sprinklers.
- High-performance glazing	Can change thermal characteristics of compartment for burning. Can impact FF access.	Moderate	Sprinklers. Assure adequate FD access. Assure mechanism for FD smoke/heat venting. Approved / listed materials.
- Low-emissivity & reflective coating	Can change thermal characteristics of compartment for burning. Can impact FF access.	Moderate	Sprinklers. Assure adequate FD access. Assure mechanism for FD smoke/heat venting. Approved / listed materials.
- Double-skin façade	Can change thermal characteristics of compartment for burning. Can impact FF access. Can present 'chimney' for vertical smoke and flame spread if not properly fire stopped.	Moderate	Appropriate fire stop between floors. Sprinklers may have some benefit (sprinklered building). Assure mechanism for FD smoke/heat venting. Approved / listed materials.
- Bamboo, other cellulosic	Can contribute to flame spread, smoke development and fuel load.	Moderate	Approved / listed materials. Flame retardant treatments. Sprinklers.
- Bio-polymers, FRPs	Can contribute to flame spread, smoke development and fuel load.	Low	Approved / listed materials. Flame retardant treatments. Sprinklers.
- Vegetative roof systems	Can contribute to fire load, spread of fire, impact FF operations, impact smoke and heat venting, contribute to stability issues.	Moderate	Manage fire risk of vegetation. Assure use of fire tested components. Provide adequate area for FD acces, smoke/heat venting, and other operations.Approved / listed materials.
- PVC rainwater catchment	Can contribute additional fuel load.	Low	Limit volume.
- Exterior cable / cable trays	Can contribute additional fuel load.	Low	Limit volume.Approved / listed materials.
Façade Attributes			
- Area of glazing	Can present more opportunity for breakage and subsequent fire spread and/or barrier to FF access depending on type.	Moderate	
- Area of combustible material	Larger area (volume) provides increased fuel load.	High	Limit volume.
- Awnings	Impacts FF access.	Low	
- Exterior vegetative covering	Can impact FF access and present WUI issue.	Low	Limit volume.

Fig. 5.3 Tabular representation of green element, hazard, concern level and mitigation

(residential and retail), BREEAM (new buildings), GREEN MARK (residential and nonresidential), and the IgCC. Review of this sample of green building rating schemes and the IgCC indicated that fire safety objectives are not explicitly considered in these systems. This is understandable since the focus is sustainability. However, it means that the potential exists for competing objectives with fire and life safety concerns.

While the sample of green building rating schemes did not include any schemes that included fire safety objectives, it was determined during the research effort that the green building rating scheme of the German Sustainable Building Council (DGNB) includes criteria for fire prevention (http://www.dgnb-system.de/dgnb-system/en/system/criteria/, accessed last on 10/29/12). However, detail on the weights of fire prevention attributes relative to the green attributes was not able to be verified. Nonetheless, it is understood that some credit is given for fire protection features such as smoke extract, automatic sprinklers, and structural fire protection. Likewise, it was determined that BREEAM-in-USE (http://www.breeam.org/page.jsp?id=373, last accessed on 10/29/12), a recent BRE scheme to help building managers reduce the running costs and improve the environmental performance of existing buildings, incorporates fire risk reduction attributes (BRE 2010). The fire risk reduction attributes related to such issues as whether a fire risk assessment has been conducted, are emergency plans in place, and so forth. No indication of consideration of fire protection systems was identified as part of BREEAM-in-USE.

Given the nature of the green building rating schemes that were reviewed, and the use of these systems by architects and others for which imagery is as illustrative as test, the approach taken for the review and presentation of these systems was to identify and pictorially classify a set of potential fire risk/hazard considerations and potential fire impacts, and to identify in each rating scheme reviewed where such risks might be introduced. The list of hazard/risk considerations, fire impacts, and consolidated representation of fire hazard and impact are illustrated in Figs. 5.4, 5.5, 5.6.

	Schematic Depiction of Interaction	Hazard Description		Schematic Depiction of Interaction	Hazard Description
1		Insulation increases interior temperature	8		The use of greenery systems on the envelope invades the surface with combustible material (dry and local species required)
2		Natural lighting and ventilation requires a non-compact building form	9		Natural ventilation requires connecting the exterior with the interior building parts
3		Renewable energy systems require invading part of the building envelope	10		Water consumption reduction influences the election of fire suppression systems
4		Daylight control devices require invading part of the building envelope	11		Noise reduction strategies require the use of non-rigid (elastic) joints
5		Disposal of rooms for waste or recyclable materials increases fire loads	12		Vegetation protection and use of greenery to reduce heat island effect influence the building surroundings conditions (may affect fire conditions)
6		Specific facade materials and systems reduce the election for optimal performance and may affect fire conditions	13		Structural materials prone to quicker failure.
7		Specific interior materials and systems reduce the election for optimal performance and may affect fire conditions	14		Adds additional fuel load to building

Fig. 5.4   Schematic representation of fire hazard

Interior Spread	Exterior Spread	Evacuation	FP Systems	Fire Service	Structure

Fig. 5.5 Schematic representation of fire impact

	Schematic Depiction of Interaction	Hazard Description	Primary Issues Associated with Fire Interaction with Green Elements					
			Interior Spread	Exterior Spread	Evacuation	FP Systems	Fire Service	Structure
1		Insulation increases interior temperature	✓	✓	✓	✓	✓	✓
2		Natural lighting and ventilation requires a non-compact building form	✓	✓	✓	✓	✓	✓
3		Renewable energy systems require invading part of the building envelope	✓	✓			✓	
4		Daylight control devices require invading part of the building envelope		✓			✓	✓
5		Disposal of rooms for waste or recyclable materials increases fire loads	✓		✓	✓		✓
6		Specific facade materials and systems reduce the election for optimal performance and may affect fire conditions		✓		✓	✓	✓
7		Specific interior materials and systems reduce the election for optimal performance and may affect fire conditions	✓		✓	✓	✓	✓
8		The use of greenery systems on the envelope invades the surface with combustible material (dry and local species required)		✓			✓	✓
9		Natural ventilation requires connecting the exterior with the interior building parts	✓	✓	✓		✓	
10		Water consumption reduction influences the election of fire suppression systems	✓		✓	✓		✓
11		Noise reduction strategies require the use of non rigid (elastic) joints	✓		✓		✓	
12		Vegetation protection and use of greenery to reduce heat island effect influence the building surroundings conditions (may affect fire conditions)		✓			✓	
13		Structural materials prone to quicker failure.	✓	✓	✓	✓	✓	✓
14		Adds additional fuel load to building	✓	✓	✓	✓	✓	✓

Fig. 5.6 Schematic representation of fire impacts associated with green building features

Categories / Chapter	Section / Assessment Issues	Aims	Credits / Scores	Requirements / Criteria	Procedures	Fire Hazard Primary (P) & Secondary (S)	Summary
Energy Efficiency (116 p. max.)	NRB 1-1 Thermal performance of Building Envelope- ETTV Enhance the overall thermal performance of the building envelope to minimize heat gain thus reducing the overall cooling load requirement		12	Maximum ETTV = 50 W/m2	Increase envelope insulation	(P) Increasing insulation (diminishing ETTV value) will increase interior temperature under fire (S) Can affect compartmentalization, the structure time resistance or evacuation	
				Use sun control devices		(P) May influence fire exterior spread (S) May disturb fire brigade intervention	
	NRB 1-2 Air-Conditioning System Encourage the use of better energy efficient air-conditioned equipment to minimize energy consumption		30	(a) Water-Cooled Chilled-Water Plant (b) Air Cooled Chilled-Water Plant/Unitary Air-Conditioners			
	NRB 1-3 Building Envelope-Design/Thermal Parameter Enhance the overall thermal performance of building envelope to minimize heat gain		35	(a) Minimum direct west facing façade through building design orientation (b) (i) Minimum west facing window openings (ii) Effective sunshading provision for windows on the west façade with minimum shading of 30% (c) Better thermal transmittance (U-value) of external west facing walls (≤ 2 W/m2K) (d) Better thermal transmittance (U-value) of roof	Increase envelope insulation	(P) Increasing insulation (diminishing ETTV value) will increase interior temperature under fire (S) Can affect compartmentalization, the structure time resistance or evacuation	

Fig. 5.7   Extract from GREEN MARK assessment for potential unintended fire consequences

Using the above system, the sample group of green building rating schemes and the IgCC were then reviewed, and the illustrative schematics for fire hazard impacts associated with green building features were applied. An illustration of the resulting comparison matrix is shown in Fig. 5.7.

For the review, each rating system/code that was reviewed was converted into the format below, identifying Category/Chapter in the document, Section/ Assessment issue being addressed, Aims of the scheme (where stated), Credits/ Scores (where provided/used in the scheme), specific Requirements/Criteria required to be achieved, Procedures to be followed in achieving the targeted performance, Primary or Secondary fire hazards that were identified as part of the review, and a Summary of the potential fire impact represented by the schematic illustration (Fig. 5.4).

It should be noted that the conversion of the various green rating schemes and the IgCC into a common format required some interpretation on the part of the project team, since different terminology is used in each system. However, the intent is to be illustrative of the potential fire hazards for similar elements across the systems, so the need to exactly reproduce each rating system was not seen as imperative. Full schematic comparisons are provided in Appendix G.

Chapter 6
Summary and Conclusions

A global literature review was undertaken to (a) identify actual incidents of fires in green buildings or involving green building elements, (b) identify issues with green building elements or features which, without mitigating strategies, increase fire risk, decrease safety or decrease building performance in comparison with conventional construction, (c) identify reports, studies and best practice cases which speak to the issue of addressing fire risk introduced by specific green building design elements, and (d) identify research studies in which building safety, life safety and fire safety have been incorporated as an explicit element in green building indices. In addition, consideration was given to how one might express the level of increased risk or hazard, or decreased performance, associated with fire performance of green building features. Steps were also taken to identify gaps and specific research needs associated with understanding and addressing fire risk and hazards with green building design.

Outcomes of this effort include the following:

- There are currently no fire incident reporting systems in the United States or other countries surveyed which specifically collect and track data on fire incidents in green buildings or on items labeled as green building elements or features. Unless changes are made to reporting systems such as NFIRS, it will be difficult to track such fire incident data.
- Web searches and surveys have identified more than two dozen reported fire incidents that are related to green issues. Examples include fires associated with photovoltaic (PV) panels and roof materials, fire and safety hazards attributed to increased energy efficiency aims in residential buildings (primarily insulation related), fire involving insulating materials, fires associated with exterior cladding that contains combustible insulation materials or coatings, and fire performance of timber frame buildings with lightweight engineered lumber (LEL) components.
- Studies related to green building and fire issues produced by BRE, BRANZ, FMGlobal and the NASFM have been identified. Research on specific building elements with green attributes, but not necessarily labeled as green, such

B. Meacham et al., *Fire Safety Challenges of Green Buildings*, SpringerBriefs in Fire, DOI: 10.1007/978-1-4614-8142-3_6, © Fire Protection Research Foundation 2012

as lightweight engineered lumber (LEL), has been identified at UL and NRC Canada.

- From the materials reviewed, a comprehensive list of green building site and design features/elements/attributes has been compiled. The list is titled "Table 3.1. Green Building/Site Elements and Attributes" and can be found in Chap. .
- From the materials reviewed, a list of fire-related hazards and risk factors, associated with green building elements, has been compiled. The list is titled "Table 4.1. Hazard, Risk and Performance Attributes" and can be found in Chap. 4.
- Using Tables 3.1 and 4.1, a set of matrices relating green attributes and potential fire hazards was developed. The matrix concept is illustrated in "Fig. 5.1. Matrix of Green Attributes and Potential Fire Hazards" and can be found in Chap. 5. The complete set of matrices is detailed in Appendix E.
- Using the matrices identified above, an approach for illustrating the relative fire risk or hazard, or decreased fire performance, associated with green building elements, was developed. The relative risk matrix is illustrated in "Fig. 5.2. Relative Fire Risk/Hazard Level of Green Attributes" and can be found in Chap. 5. A complete matrix, which is based on a qualitative expert judgment approach for illustrating relative risk levels, can be found in Appendix F.
- Potential mitigation strategies for addressing the relative increase in fire risk or hazard associated with the green building elements and features have been identified. These are presented at a basic level (e.g., provide automatic sprinkler protection). In many cases, adherence with existing test standards, codes and related design guidelines associated with conventional construction will help mitigate potential increases in fire risk or hazard associated with green building elements. Approval or certification of products which meet loss prevention criteria, and are indicated as having some type of green attribute which might gain credit in a green certification scheme, have been identified within the FM Approvals system and UL Product Certification system.
- Review of a sample of green rating schemes for which data were readily available, including LEED (residential and retail), BREEAM (new buildings), GREEN MARK (residential and nonresidential), as well as the IgCC, indicates that fire safety objectives are not explicitly considered. However, as noted above, implementation of certain green features could have a negative impact on fire or life safety if not mitigated. A qualitative approach, using text and pictograms, was used to reflect areas of fire and life safety concern, as illustrated in "Fig. 5.4. Fire Hazards with Green Building Features" and "Fig. 5.5. Extract from GREEN MARK Assessment for Potential Unintended Fire Consequences", which can be found in Chap. 5. Detailed matrices of the assessment of the green rating schemes for potential unintended fire consequences can be found in Appendix G.
- It was determined that the green building rating scheme of the German Sustainable Building Council (DGNB) includes criteria for fire prevention (http://www.dgnb-system.de/dgnb-system/en/system/criteria/, accessed last on 10/29/12). Detail on the weights of fire prevention attributes relative to the green attributes was not able to be verified; however, it is understood that some credit is given for fire protection features such as smoke extract, automatic sprinklers, and structural fire protection.

- It was determined that BREEAM-in-USE (http://www.breeam.org/page.jsp?id=373, last accessed on 10/29/12), a recent BRE scheme to help building managers reduce the running costs and improve the environmental performance of existing buildings, incorporates fire risk reduction attributes. The fire risk reduction attributes related to such issues as whether a fire risk assessment has been conducted, are emergency plans in place, and so forth. No indication of consideration of fire protection systems was identified.

In order to fill gaps in knowledge to better address fire issues with green building features, further research is suggested in several areas.

- To address the lack of reported fire experience with green buildings and green building elements, especially in buildings which have a green rating or certification, a modification is required to fire incident data reporting systems as NFIRS. This could perhaps be done in collaboration with the USGBC and/or AIA, and parallel organizations in other countries. If this avenue is pursued, there will be challenges associated with how responding fire departments are able to identify LEED, BREEAM, or other such ratings for buildings. In cases where ratings or certifications are posted on buildings (e.g., LEED, Energy Star, BREEAM or other), this information could be readily captured by fire fighters responding to fire incidents in the building. In cases where such ratings or certifications are not posted, inclusion of specific features such as 'double-wall façade' or 'LEL' might be added to the incident reporting system, or additional guidance can be provided to first responders in identifying green attributes of buildings.
- To address the lack of analysis on fire 'risk' associated with green building elements, it is suggested that a more extensive research project is needed to review existing studies and reports on fire performance of green building elements, even if not explicitly identified as such (e.g., LEL). Research is needed to (a) develop a clear set of comparative performance data between green and 'conventional' methods, (b), develop an approach to convert the relative performance data into relative risk or hazard measures, and (c) conduct a risk (or hazard) characterization and ranking exercise, with a representative group of stakeholders, to develop agreed risk/hazard/performance levels.
- To explore the extent to which current standard test methods are appropriate for evaluating both green and fire safety criteria, and result in adequate mitigation of fire risk/hazard concerns, investigation into level of fire performance delivered by current standard test methods and into the in situ fire performance of green building elements is recommended.
- To address the lack of published case studies in which increased fire risk or hazards associated with green building elements have been specifically addressed, groups such as SFPE, NFPA, AIA and the USGBC can be encouraged to hold symposia on these topics and encourage publication of case studies in proceedings and associated journals. While such some studies have been published, they mostly reflect 'issues or concerns' with green building features without significant quantification of impacts and formal risk analysis.

- To address the lack of studies which have investigated incorporating building safety, life safety and fire safety as explicit elements in green building indices, joint research efforts between the FPRF and the USGBC and other promulgators of such indices could be explored with the aim to incorporate fire and life safety objectives as fundamental elements in green rating schemes and codes.
- To facilitate better collection of relevant data on fire safety challenges with green buildings in the future, a fire and green building data repository could be established. This might build on an existing effort (e.g., http://www.firemarshals.org/programs/greenbuildingsandfiresafetyprojects.html) or be supported by the FPRF or other organizations.

Appendix A
Informational Resources

This section provides a listing of resources identified during the search for information on fire incidents in green buildings, reports on fire studies of green building elements, and related information. While not all items were referenced or were ultimately of value to the project, they are included for completeness. Specific resources listed and discussed in Chap. 4 and other sections of this report are not necessarily repeated here.

It is worth highlighting that the following resources provide significant discussion relative to the project focus and are highly recommended as key sources of information on the topic of green buildings and fire:

The BRANZ study, *Building Sustainability and Fire-Safety Design Interactions*, http://www.branz.co.nz/cms_show_download.php?id=716733515027fe462618 8881f674635d51e3cfb0 (last accessed 10/21/12)

The BRE study, *Impact of Fire on the Environment and Building Sustainability*, http://www.communities.gov.uk/documents/planningandbuilding/pdf/1795639.pdf (last accessed 10/21/12)

The NASFM Green Buildings and Fire Safety Project (report and web links), http://www.firemarshals.org/programs/greenbuildingsandfiresafetyprojects.html (last accessed 10/21/12)

Conference, Trade Magazine, Journal Articles and Government Reports

Barber, D. (2012). "Can Building Sustainability be Enhanced through Fire Engineering of Structures?," *Proceedings, 9th International Conference on Performance-Based Codes and Fire Safety Design*, SFPE, Bethesda, MA. Discussion of fire and sustainability concepts, including what makes a material 'sustainable.'

Bill, R.G., Jr., Meredith, K., Krishnamoorthy, N., Dorofeev, S., and Gritzo, L.A. (2010), "The Relationship of Sustainability to Flammability of Construction Material," 12th International Fire Science & Engineering Conference, *Interflam 2010*, Interscience Communications, Ltd. Discusses flammability of construction materials which have been identified as having a green or sustainable attribute.

B. Meacham et al., *Fire Safety Challenges of Green Buildings*, SpringerBriefs in Fire, DOI: 10.1007/978-1-4614-8142-3, © Fire Protection Research Foundation 2012

BRE 2709 (2010). *Impact of fire on the environment and building sustainability BD 2709*. Department for Communities and Local Government, UK. Last accessed 10/21/2012. Available at http://www.communities.gov.uk/documents/p lanningandbuilding/pdf/1795639.pdf

Carter, M., Lee, N., Oliver, E. & Post, M. 2011. *Promoting the Design of Buildings that are Fire Safe and Sustainable, A Review for Fire Protection Association Australia*. An Interactive Qualifying Project Report submitted in Partial Fulfillment of the Bachelor of Science, Worcester Polytechnic Institute, Worcester, MA

Charters, D.A. and Fraser-Mitchell, J. (2007). "The Potential Role and Contributions of Fire Safety to Sustainable Buildings," *Interflam 2007*, Interscience Communications Ltd., pp, 1231–1242. Discusses contribution of fire protection systems and features to building sustainability.

Chow, C.L., Chow, W.K. (2003). "Assessing Fire Safety Provisions for Satisfying Green or Sustainable Building Design Criteria: Preliminary Suggestions," *International Journal on Architectural Science*, 4(3):141,142–146. Academic study on assessment of fire safety provisions for satisfying green or sustainable building design criteria and preliminary suggestions. Recommends creation of assessment scheme for inspecting fire safety in green or sustainable buildings.

Chow, W.K. & Hung, W. Y. (2006). Effect of cavity depth on smoke spreading of double-skin facade. *Building and Environment*, 41, 970–979. Discusses testing relative to smoke propagation within cavity of double-skin façade. Shows relationship based on depth, with 1 m allowing for smoke spread in the testing that was conducted.

Chow, W.K. and Chow, C.L. (2005). "Evacuation with Smoke Control for Atria in Green and Sustainable Buildings," *Building and Environment*, 40, pp. 195–200. Discusses competing objectives of building, fire and green codes and attributes, focusing on how atria—which meet green criteria but not all prescriptive fire codes in Hong Kong—can aid evacuation during fire.

Chow, W.K., Hung, W.Y., Gao, Y., Zou, G. and Dong, H. (2007). "Experimental Study on Smoke Movement Leading to Glass Damages in Double-Skinned Façade," *Construction and Building Materials*, 21, pp. 556–566. Discusses testing conducted relative to breaking glass on double-skin façade due to fire. Testing showed cavity depth of 1 m resulted in glass breakage and smoke spread concerns.

DECC (2010). The green deal: A summary of the Government's proposals. Department of Energy and Climate Change, England. Report number 10D/996. Available at http://www.decc.gov.uk/assets/decc/legislation/energybill/1010-green-deal-summary-proposals.pdf Accessed 6/26/2012. UK's proposal of the Green Deal: what it is and why it's needed.

Dent, S. (2010). "Fire protection engineering and sustainable design why performance-based design will become increasingly important in the future," *Fire Protection Engineering*, Penton. Available from: http://magazine.sfpe.org/fire-protection-design/fire-protection-engineering-and-sustainable-design. Accessed 6/26/2012. Discusses that performance-based design will become increasingly important in the future due to issues associated with the challenges a fire protection engineer faces with sustainable/green building design and increased number of people inside a building. Also provides two case studies of how fire protection design worked with a green sustainable building.

Ding, W., Hasemi, Y. & Yamada, T. (2005). Natural ventilation performance of a double-skin facade with a solar chimney. *Energy and Buildings*, 37, 411–418. Discusses non-fire ventilation performance of double-skin façade.

Emberley, R.L., Pearson, T.M., Andrews, A.S. (2012) Green Buildings and Fire Performance, Research report submitted as part of Course FP571, Performance-Based Design, Worcester Polytechnic Institute, Worcester, MA. Summarizes areas of potential concern based on review of literature.

Grant, C.C. (2010). *Fire fighter safety and emergency response for solar power systems*. A report of the Fire Protection Research Foundation, Quincy, MA. Accessible from the internet at http://www.nfpa.org/assets/files/pdf/research/fftacticssolarpower.pdf. Accessed 6/26/2012. Investigation of fire fighter safety and emergency response for solar power systems. Report focuses on the fact of growing demand for alternative energy and the number of areas of concern with hazard mitigation and emergency response. Includes several case studies of fire incidents, review of fire fighter tactics, and recommendations for further research.

Green Building Challenges for the Fire Service, Fire Engineering (web article), c2011 Available from: http://www.fireengineering.com/articles/2011/01/green-building-challenges-for-the-fire-service.html. Accessed 8/1/2012. This website contains the report of Green Building Challenges for the Fire Service. Discusses several issues that face the fire service and provide justification of why it is a problem.: Site and Landscape Issues, Building Envelope, Vegetative roof systems, high performance glazing, building design attributes, water conservation, alternative power systems, wind turbine systems, hydrogen fuel cell power systems, battery storage systems.

Gritzo, L. A., Doerr, W., Bill, R., Ali, H., Nong, S. & Kranser, L. (2009). *The Influence of Risk Factors on Sustainable Development*. Norwood, MA, USA: FM Global, Research Division. Available at http://www.fmglobal.com/assets/pdf/P09104a.pdf. Last accessed 10/21/2012.

Hirschler, M.M. (2008). "Polyurethane foam and fire safety," *Polym. Adv. Technol.*; 19: 521–529. Includes discussion of toxic impacts of flame retardants.

Hofmeister, C. E. (2010). "Prescriptive to Performance-Based Design in Green Buildings," *Fire Protection Engineering*, Penton Press. http://fpemag.com/archives/article.asp?issue_id=54&i=452 Accessed 3/24/2012

Jarrett, R., Lin, X.G., and Westcott, M. (2011) *CSIRO risk profile analysis—guidance for the home insulation safety program*. CSIRO Report No. EP112079. Australia. The 2011 report for the Home Insulation Safety Program that set out to give advice on the Home Insulation Program and identify and rank dwellings with insulation according to risk indicators. Report provides information of Fire Risk, Safety Risk, Inspection Results, and Risk Profiling Tool.

Kasmauskas, D. G. (2010). "Green Construction and Fire Protection: Will LEED eventually recognize the environmental benefits of fire sprinklers?," *Fire Protection Engineering*, Penton Press. http://magazine.sfpe.org/fire-protection-design/green-construction-and-fire-protection Accessed 3/24/2012.

Kortt, M.A. and Dollery, B. (2012). "The home insulation program: An example of Australian government failure," *The Australian Journal of Public Administration*; Vol. 71, No. 1, pp. 65–75 doi:10.1111/j.1467-8500.2012.00754.x. Report on the failure of the Home Insulation Program. Offers start to finish information of program including analysis and lessons for the future.

McVay, P., Jackson, K., Christian, M., Fitzgerald, W., Jognson. M, Hall, M. and Cass, B. (2010). *Home Insulation Program*, ANAO Audit Report No.12 2010–2011. Australia: Australian National Audit Office. Audit report on the Home Insulation Program that details why the program went into place, the failing of the Home Insulation Program, creation of the Home Insulation Safety Program, and key milestones of the program as well (what phase the program was in and money used to that point.)

Miller, L., Joyce, S., and Wamakima, D. (2008). *Fire Safety in Green Buildings*. Interactive Qualifying Project, Worcester Polytechnic Institute, Worcester, MA.

Modern Green-Building Fire Protection [Internet]: HPAC Engineering; c2009 (cited 2012 August 1). Available from: http://hpac.com/fastrack/Modern-Green-Building-Fire-Protection/. Encourages the use of fire safety and fire protection practices through use of awareness, codes and standards, LEED certification, fire detection technology, and fire suppression technology.

Murphy, J. J. and Tidwell, J. (2010). *Bridging the Gap*: Fire Safety and the Green Building, National Association of State Fire Marshals.http://www.firemarshals.org/greenbuilding/bridgingthegap.html Last accessed 8/12/2012.

Ohlemiller, T.J. and Shields, J.R. (2008) Aspects of the Fire Behavior of Thermoplastic Materials, NIST Technical Note 1493, Gaithersburg, MD.

Robbins, A.P. (2012). *Building Sustainability and Fire-Safety Design Interactions*: Scoping Study, BRANZ Study Report 269. Available at http://www.branz.co.nz/cms_show_download.php?id=716733515027fe4626188881f674635d51e3cfb0. Last accessed 10/21/2012.

Shields, T. J., Silcock, G. W. H. & Hassani, S. K. S. (1997). The behavior of double glazing in an enclosure fire. *Journal of Applied Fire Science*, 7, 267–286.

Spadafora, R.R. (2009). The fire service and green building construction: An overview. Fire Engineering.

Starr, S. (2010). Turbine Fire Protection. *Wind Systems* 2010 (August):44, 45–51.

Stec, W. J. & Van Paassen, A. H. C. (2002). Validation of the simulation models of the double skin facade. *Advances in Building Technology*, 2, 1181–1188.

Wieczorek, C., Ditch, B. and Bill, R. (2010). *Environmental Impact of Automatic Sprinklers*, FMGlobal Technical Report. Available at http://www.fmglobal.com/assets/pdf/P10062.pdf. Last accessed on 10/21/2012 (registration may be necessary).

Informational Articles/Resources about Fire and Green Elements/Systems

Summary of Wind Turbine Accident data to 30 June 2012

Caithness Windfarm Information Forum [Internet]: Caithness Windfarm; c2012 [cited 2012 August 1]. Available from: http://www.caithnesswindfarms.co.uk/accidents.pdf. This is a document by Caithness Windfarm of Wind Turbine

Accidents to 30 June 2012. The document has a long list of the different types of accidents that can happen when using a Wind Turbine. It lists fire as the second most common accident and the difficulty in extinguishing them.

Do HVLS Fans Risk Fire Safety? [Internet]: Klausbruckner and Associates; c2011 [cited 2012 August 1]. Available from: http://www.klausbruckner.com/blog/do-hvls-fans-risk-fire-safety/. This article contains information about a recent study of HVLS fans. The study focused on: Do HVLS fans obstruct sprinkler operation in case of fire and does their additional air flow produced while in operation increase fire spread or negatively impact overall fire dynamics?

Fire & Flammability [Internet]: Solar America Board for Codes and Standards; c2011 [cited 2012 August 1]. Available from: http://www.solarabcs.org/current-issues/fire.html. This website contains a list of Codes and Standards identified fire and flammability safety areas. It has links to articles of : "The Ground-Fault Protection: 'Blind Spot' A Safety Concern for Larger PV Systems in the U.S.," "Prevention of PV Modules as the Cause of a Fire," "Fire Class Rating of PV Systems" and "Fire Fighter Safety in Buildings with PV Modules."

Fire-Resistant Insulation and Systems [Internet]Webpage: This Old House [cited 2012 6/29]. Available from: http://www.thisoldhouse.com/toh/photos/0,,20156142_2036 2077,00.html. Popular Home improvement magazine that lists Fire safety items to install in homes. Includes some factoids on wool insulation fire safety rating.

Fire Code Requirements for Spf Applications [Internet]: Sprayfoam.com [cited 2012 August 1]. Available from: http://www.sprayfoam.com/leps/letitem.cfm?letid=38. Industry website that contains information of fire code requirements (NFPA, International Residential Code, ASTM, ICC) for spray foam applications. Specific data includes Fire Test Requirements, Surface Burning Characteristics, Special Approvals, Exceptions to the Thermal Barrier Rule, Ignition Barriers, Steiner Tunnel Test and procedure for it, Foam Plastics and the FTC Consent Decree, Room Corner Fire Tests, Thermal barrier test and Hourly Fire Ratings, Hourly Fire Ratings, Thermal Barrier Test, Attic and Crawl Space Test, Standard Fire Test for Evaluation of Fire Propagation Characteristics of Exterior Non-Load Bearing Wall Assemblies Containing Combustible Components, Evaluations Services and Reports, and Acceptance Criteria for Spray Polyurethane Foam.

FAQ Radiant Barrier & Cellular Bubble Foil Insulation [Internet]: Insulation STOP.com [cited 2012 August]

Underwood J. Fire-resistant details studying the houses that survived the 1993 Laguna beach fire storm yields lessons in building to withstand the heat. Fine Homebuilding.

Standards for green Building Components

ANSI/SPRI VF-1 External Fire Design Standard for Vegetative Roofs. ANSI, Green Roofs for Healthy Cities,; 2010 January 29, 2010:http://www.spri.or g/pdf/ansi_spri_vf-1_external_fire_design_standard_for_vegetative_roofs_ jan_2010.pdf. Codes and Standards listed by the Approved American National Standard and made for Green Roofs for Healthy Cities. Contains requirements for Vegetative Roofs including definitions, system requirements & general design considerations, vegetative roof design options and maintenance.

Approval Standard for Vegetative Roof Systems [Internet]: FM Approvals; c2010 [cited 2012 August 1]. Available from: http://www.fmglobal.com/assets/pdf/fm approvals/4477.pdf. Standard for Vegetative Roof System that are used within an FM Approved roof assembly. Contains General information, general requirements, performance requirements, and operations requirements.

ANSI/SPRI VF-1 External Fire Design Standard for Vegetative Roofs [Internet]: Green Roofs for Healthy Cities; c2010 [cited 2012 August 1]. Available from: http://www.greenroofs.org/resources/ANSI_SPRI_VF_1_Extrernal_Fire_Design_Standard_for_Vegetative_Roofs_Jan_2010.pdf. Codes and Standards listed by the Approved American National Standard and made for Green Roofs for Healthy Cities. Contains requirements for Vegetative Roofs including definitions, system requirements & general design considerations, vegetative roof design options and maintenance.

ICC-700 2008. National Green Building Standard. Washington, DC: International Codes Council.

BREEAM Homepage, The world's leading design and assessment method for sustainable buildings [Online]. Watford, UK: BRE Global. Available: http://www.breeam.org/

GREEN MARK, BCA, Singapore, http://www.bca.gov.sg/greenmark/green_mark_buildings.html

IgCC, International Green Construction Code, International Code Council.

LEED 2009 For Retail: New Construction and Major Renovations. USGBC; 2009 March 2010. Codes and Standards listed by the USGBC. Contains requirements for LEED certification For Retail: New Construction and Major Renovations. Specifies 7 different categories in which a retail structure can earn points towards LEED certification for having certain green elements.

LEED for homes rating system. USGBC; 2008 January 2008. Codes and Standards listed by the USGBC. Contains requirements for LEED certification For Homes. Specifies 7 different categories in which a home can earn points towards LEED certification for having certain green elements.

Articles/Websites Related to Fire Associated with Australian Home Insulation Plan

Australian government's home insulation safety plan. Australian Government News. May 18. 2012. This is an article published on the Newswire by the Australian Government News. The article details general information about the Australian Government's Home Insulation Safety Plan including its set completion date and how many homes that have been inspected by the plan.

Foiled: Garret bans 'dangerous' roof insulation [Internet]Brisbane Times.com. au: Brisbane Times; cFebruary 9, 2010 [cited 2012 5/29]. Available from: http://www.brisbanetimes.com.au/queensland/foiled-garrett-bans-dangerous-roof-insulation-20100209-nojb.html. This is an article published by Brisbane Times in Australia. The article details about the Federal Environment Minister Peter Garrett ban on the use of foil insulation. The article continues offering details of why foil insulation was banned citing electrical problems and deaths during insulation.

Metal foil insulation banned from Federal Government's roof insulation program after four deaths [Internet] The Daily Telegraph AAP: The Daily Telegraph; cFebruary 10, 2010 [cited 2012 5/29]. Available from: http://www.dailytelegraph.com.au/property/metal-foil-insulation-banned-from-federal-governments-roof-insulation-program-after-four-deaths/story-e6frezt0-1225828479206. This is an article published by the Telegraph in Australia. The article details about Federal Environment Minister Peter Garrett ban on the use of foil insulation. The article continues offering details of affected houses, industry, and critics opinions on the subject.

Foil insulation banned amid electrocution fears [Internet] ABC News Online. AU: ABC News; cFebruary 09, 2010 12:57 [cited 2012 5/29]. Available from: http://www.abc.net.au/news/2010-02-09/foil-insulation-banned-amid-electrocution-fears/324984. This is an article published by ABC News. The article details about Federal Environment Minister Peter Garrett ban on the use of foil insulation. The article contains quotes from Peter Garret's statement to the press.

3 out of 4 products do not meet flammability standards [Internet]Kingspan Insulation- Australia: Kingspan; c6/04/2010 [cited 2012 5/29]. Available from: http://www.kingspaninsulation.com.au/News/2010/3-out-of-4-products-do-not-meet-flammability-stand.aspx. Kingspan Insulation is an Australia based company that prides itself in innovative thermal insulation products. In one of their news releases, they detail a study that 3 out of 4 products (foil insulation products) do not meet a flammability index specified by the Building Code of Australia.

Key Statistics [Internet]: Australian Government; c2012 [cited 2012 August 1]. Available from: http://www.climatechange.gov.au/government/initiatives/hisp/key-statistics.aspx. This government website contains key statistics on the Home Insulation Safety Plan. Specifically, it offers the total number of inspections done by the Plan (foil, non-foil), fire incidents linked to the HIP, and calls to the safety hotline.

Study reveals shortfall in insulation standard [Internet]: Kingspan; c2008 [cited 2012 August 1]. Available from: http://www.kingspaninsulation.asia/News/2008/Study-reveals-shortfall-in-insulation-standard.aspx. Trade website newsletter, Kingspan Australia, details a recent study by University of South Australia discovered the R value of foil-backed glass wool building blanket in-situ was up to 60 % less than the certified R-value.

A foil to the insulation debate [Internet]ABC News Blog: ABC News; c11 Feb 2010 [cited 2012 5/29]. Available from: http://www.abc.net.au/environment/articles/2010/02/11/2816444.htm. Blog article from ABC Australia summarizing the what has gone wrong with the Home Insulation Program. Article sides on poor insulation for the majority of the problems that have occurred by the HIP.

Reflective Insulation [Internet]Australian Government Insulation Rebates: Australian Government [cited 2012 5/29]. Available from: http://australian-government-insulation-rebates.com/Products/reflective-insulation.html. Australian Government Insulation Rebate website with details on the HIP.

CSIRO report- FACT SHEET. Australia: CSIRO. Condensed version of the CSIRO Report. Contains information on how many homes were installed under

HIP, and expected fire incident rate for households with insulation before, after, and without insulation.

Bita N. Unsafe batts cost $273m to redress. AUS May 18, 2012(1-All-round Country Edition). News article that talks about the amount of tax payer money used to correct the Home Insulation Programs failure. Quotes politicians calling the insulation scheme " one of the most disastrous government programs in living history."

Belukso M, Bruno F. Installation survey and thermal testing of continuous roll form foil Backed fiberglass building blanket insulation. Australia: Institute for Sustainable Systems and Technologies; 2008 22 August 2008. Report nr AFIA 12082008 v2. University Study that was cited by Kingspan that studies installation and thermal testing of continuous roll form foil backed fiberglass building blanket insulation. Report found that the R value can be up to 41 to 60 % lower than the certified R value.

Articles about Green Issues, Green Programs and Building Products in General

Energy Saving Products [Internet]Website: British Gas; c2012 [cited 2012 June 26]. Available from: http://www.britishgas.co.uk/products-and-services/energy-saving.html. This website details information about British Gas and its options in The Green Deal UK. There are several tabs of information detailing the effectiveness of insulation and energy saving products.

Only one in five consumers will use Green Deal, survey finds [Internet] Website: GreenWise; c2012 [cited 2012 June 26]. Available from: http://www.greenwisebusiness.co.uk/news/only-one-in-five-consumers-will-use-green-deal-survey-finds-3405.aspx. This article contains information about the Green Deal and a recent survey published by the BSI. The survey interviewed people on their knowledge about the Green Deal and other green products. One factoid shows that ~90 % of the 1,200 respondents felt like they did not know enough about the Green Deal.

Mason R. Coalition's green deal plans to insulate 14 million homes 'spiralling out of control.' The Telegraph (http://www.telegraph.co.uk/journalists/rowena-mason/9320404/Coalitions-Green-Deal-plans-to-insulate-14-million-homes-spiralling-out-of-control.html) .Accessed 8/24/2012. News article that contains information of how UK's Green Deal is not going as planned. Sources that companies that did sponsor the Green Deal are now holding off development further.

Millard E. Greening up: Rooftop farms and gardens flourish in the cities. MinnPost 2011 07/15/2011.

Ray SD. Energy saving potential of various roof technologies. Massachusetts Institute of Technology: Massachusetts Institute of Technology; 2009 May 05 2010.

Shonnard DR, Kicherer A, Saling P. Industrial applications using BASF eco-efficiency analysis: perspectives on green engineering principles. Environ Sci Technol 2003; 37(23):5340, 5341–5348.

Mass Save [Internet]; c2012 [cited 2012 August 1]. Available from: http://www.masssave.com/about-mass-save. This government website contains information about the Mass Save plan. Specifically it details specific information for "Your Home," "Your Business" and "For Industry Professionals" to get involved in the green movement.

Home Weatherization [Internet]: Energy.gov; c2012 [cited 2012 August 1]. Available from: http://energy.gov/public-services/homes/home-weatherization. This government website contains information on how to Weatherize your home. It details about the different types of insulation available, how insulation works, among other facts.

BSI-052: Seeing Red Over Green Roofs [Internet]: Building Science; c2011 [cited 2012 August 1]. Available from: http://www.buildingscience.com/documents/insights/bsi-052-seeing-red-over-green-roofs/. This is an opinionated article against the use of Green Roofs. The author provides facts of the dangers of green roofs in a detailed manner of how they work. He also points out the current absurdity of using green technology is woefully energy inefficient compared to the other options out there.

Insulation: are there cracks in the batts story? [Internet]: Habitech Systems; c2011 [cited 2012 August 1]. Available from: http://www.habitechsystems.com.au/news/2011/5/9/insulation-are-there-cracks-in-the-batts-story.html. This is an industry website detailing information about Insulation. Specifically the website points out the difficulty in installing insulation and the effectiveness of insulation batts.

Cellulose insulation vs. foam insulation [Internet]: Tascon Industries, Inc.; c2011 [cited 2012 August 1]. Available from: http://www.tasconindustries.com/CelluloseVsFoam.html#homefires. This is an industry website detailing information about cellulose insulation vs. foam insulation. The website provides a quick comparison table of the different characteristics between types of insulation. It also contains an article about home fires linked to spray-foam installation.

Therma Fiber Commercial Insulation [Internet] [cited 2012 August 1]. Available from: http://www.thermafiber.com/InsulationProducts/CommercialInsulation. Trade website, Thermafiber, that offers information on different types of commercial insulation. Thermafiber products also contribute 33 LEED credits across 4 categories if installed.

Heating, Ventilation and Air-Conditioning (HVAC) Systems [Internet]: EPA [cited 2012 August 1]. Available from: http://www.epa.gov/iaq/schooldesign/hvac.html. Government website, EPA, that offers information on Heating, Ventilation and Air-Conditioning Systems through codes and standards, definitions, and minimum requirements.

Vital Signs [Internet]: Berkeley [cited 2012 August 1]. Available from: http://arch.ced.berkeley.edu/vitalsigns/res/downloads/rp/glazing/glaz3-bg.pdf. University Project at Berkeley that investigates High Performance Glazing through the codes and standards by state and area. Includes table where each states ability to Save Energy on Residential Energy Building Codes is graded on a report card.

Reducing Urban Heat Islands: Compendium of Strategies [Internet]: EPA [cited 2012 August 1]. Available from: http://www.epa.gov/hiri/resources/pdf/GreenRoofsCompendium.pdf. Government report on Reducing Urban Heat Island: Compendium of Strategies. Includes information on what are green roofs, how they work, the different types of green roofs, benefits and costs, other factors to consider, green roof initiatives, and last resources one could use.

MOHURD Codes and Standards- Wood [Internet] [cited 2012 August 1]. Available from: http://cn.europeanwood.org/fileadmin/ewi/media/building-with-wood-c7.pdf. Industry Document for Codes and Standards of the wood construction code system. Has list of codes and standards that are commonly used in China.

Wood-Frame Housing-A North American Marvel [Internet]: CWA [cited 2012 August 1]. Available from: http://www.cwc.ca/documents/durability/BP4_Wo odFrameHousing.pdf . Industry document that contains information about the challenges and effectiveness of using wood construction. Cites information that wood can keep energy bills down, resistant to earthquakes, resistant to storms, stands up to time, and that wood returns structural integrity during a fire.

Wood-frame Construction, Fire Resistance and Sound Transmission [Internet]: Forintek Canada Corp., Société d'habitation du Québec and Canada Mortgage and Housing Corporation. [cited 2012 August 1]. Available from: http://www.forintek.ca/public/pdf/Public_Information/fact%20sheets/Fire-Sound_ENGLISH%20FINAL.pdf. Industry document that expands upon the usefulness of wood. Includes information of how wood based products provide fire resistance and sound transmission along with pictures of how items are installed.

Details for Conventional Wood Frame Construction [Internet]: American Wood Council [cited 2012 August 1]. Available from: http://www.awc.org/pdf/WCD1-300.pdf. Industry document that details conventional wood frame construction. Contains information on how a wood frame is constructed from what wood is used to flooring. On Page 8, contains details of Firestopping.

Windows and Glazing [Internet]: WBDG; c2010 [cited 2012 August 1]. Available from: http://www.wbdg.org/resources/windows.php. Industry website that contains information about Windows and Glazing from Whole Building design Guide. Gives detailed information of how windows work, how to specify windows and glazing, representative glass specifications, other attributes, and opportunities and cautions. Also includes several case studies and links to standards and codes.

FAQ — Frequently Asked Questions about fire-rated glass & framing [Internet]: Fireglass; c2012 [cited 2012 6/29]. Available from: http://www.fireglass.com/faq/. This industry website (TGP Fire Rated) provides a FAQ about fire-rated glass and framing. The questions and answers detail information about different types of fire rated glass. Specific questions include: What are my primary options in fire-rated glass? And Why is the "fire hose stream" (thermal shock) test so important?

Appendix B
Representative Fire Issues
and Mitigation Approaches

This appendix highlights a selection of potential fire concerns with green building elements and potential, but not exhaustive, mitigation strategies. This section is not intended to be a comprehensive treatment; rather, and illustration of challenges and potential mitigation opportunities. Information was obtained during the information search from the sources as indicated.

Spray-on Foam Insulation

Issue

As reflected in the literature review, a number of fires have been reported associated with spray on foam insulation. For example:

Fire investigators suspect that a fire that destroyed a $5 million home in Woods Hole, Mass., was ignited when excess heat was generated by the exothermic reaction that occurs during the installation of spray polyurethane foam [Photo credit: Dave Curran]—http://www.greenbuildingadvisor.com/blogs/dept/green-building-news/three-massachusetts-home-fires-linked-spray-foam-installation (accessed 6/26/2012)

B. Meacham et al., *Fire Safety Challenges of Green Buildings*, SpringerBriefs in Fire,
DOI: 10.1007/978-1-4614-8142-3, © Fire Protection Research Foundation 2012

Representative Response

In response to a series of at least three fires of this type in Massachusetts, Massachusetts State Fire Marshal Stephen Coan issued a memo with the text indicated below on 1 July 2011 (http://www.mass.gov/eopss/docs/dfs/osfm/advisories/2011/20110701-spray-foam-insulation-fires.pdf):

"Recently, the Department of Fire Services, Division of Fire Safety, has become aware of a number of fires involving commercially available spray-on foam insulation. At least 3 fires, one being a fatal fire, are believed to have been started during the application of spray foam insulation, and currently remain under investigation. These foam insulation products are being increasingly utilized as part of the "green" building movement. The insulation is a two-part spray foam product and come in several types, either in a "closed" cell foam design (rigid type, solid) and "open" cell foam design (sponge like, not rigid). Information gathered by the Division of Fire Safety from different manufacturers indicate that there are several possible scenarios that could lead to a heat build-up, and a possible fire scenario. These are: improper application techniques (excessive thickness, or spraying new material into the already applied rising foam) and/or improper mixtures of the chemicals at the application nozzle."

Potential Mitigation Strategies

In this case, fire hazards appear to be concentrated during installation/application of the foam. Mitigation strategies include control of ignition sources during application, adherence to installation procedures and guidelines, and implementation of fire watch during application.

Photovoltaic Panels
Issue
As reflected in the literature review and the international survey, a number of fires have been reported associated with photovoltaic panels used of local energy and heating. The material below reflects industry perspectives on the hazards and steps taken towards mitigation http://solarjuice.com/blog/buildings-and-pv/solar-panels-and-fire/.

Representative Response/Potential Mitigation Strategies

The following material is reprinted from Solar America Board for Codes and Standards, downloaded from http://www.solarabcs.org/current-issues/fire.html (accessed 7/24/2012).

Fire Class Rating of PV Systems

Solar ABCs research investigates whether and how the presence of stand-off mounted PV arrays may affect the fire class rating of common roof covering materials. In particular, these tests were initiated in response to questions from stakeholders about the language in the UL Guide Card that stated that PV modules may or may not reduce the fire class rating of roof coverings when modules of a lower rating are installed above a roof covering with a higher rating. All tests were conducted by UL in Northbrook, IL, with assistance from representatives of Solar ABCs.

In April 2010, Solar ABCs published an interim report, *Flammability Testing of Standard Roofing Products in the Presence of Stand-off Mounted PV Modules.*

In December 2010, the Standard Technical Panel for UL Standard 1703 appointed a task group to develop a new system fire classification test. This system test will replace the current module fire classification test and will provide a better test for the impact of the photovoltaic system on the fire classification rating of the roof assembly. This task group has developed several drafts of new standard language and is working with stakeholders to obtain broad input into the proposal.

Testing

Since 2009, Underwrites Laboratories, with the support of Solar ABCs, has conducted research testing on issues related to the fire classification rating of PV modules and systems. The following reports detail the tests and results.

Effect of Roof-Mounted Photovoltaic Modules on the Flammability of Roofing Assemblies, September 30, 2009, Revised March 5, 2010

This initial study measured the surface temperature and incident heat flux of a noncombustible roof with a noncombustible PV module surrogate installed at 10, 5, and 2.5 inches above the roof. In addition, limited burning brand and spread of flame tests were conducted using actual PV modules. These tests were designed to (1) develop baseline data on the fire exposure during standard tests for roof with no PV module according to UL 170, (2) determine the effect of varying selected PV installation parameters, and (3) document the impact of lesser fire rated PV modules on common roofing assemblies.

Effect of Rack Mounted Photovoltaic Modules on the Flammability of Roofing Assemblies—Demonstration of Mitigation Concepts, September 30, 2009, Revised February 10, 2010

In a continuation of the first study, several simple design concepts were devised to assess their effectiveness in improving the fire classification rating of the roof with a rack mounted photovoltaic module. The mitigation measures studied includes (1) use of flashing at the leading edge of the roof with control of separation between the roof and flashing, and (2) use of non-combustible back sheet.

Effect of Rack Mounted Photovoltaic Modules on the Fire Classification Rating of Roofing Assemblies, January 30, 2012

The second project further investigated rack mounted PV modules on roof decks to determine (1) the effect of PV modules mounted at angles (positive and negative) to steep and low sloped roof, (2) the impact of PV modules mounted at zero clearance to the roof surface and with the ignition source directed in the plane of the roof or the plane of the PV surface, and (3) the heat release rate and transfer to roof surface of Class A, B, C brands and common materials such as leaf debris and excelsior (wood wool).

Characterization of Photovoltaic Materials — Critical Flux for Ignition/Propagation, January 16, 2012.

The third project investigated the critical flux for ignition of roofing and PV products. While the individual values varied, most were within the range of the flux values measured on the roof in the original experiments with the PV module in place.

Determination of Effectiveness of Minimum Gap and Flashing for Rack Mounted Photovoltaic Modules, March 29, 2012.

The fourth project was undertaken to validate the performance of two approaches thought to mitigate the effect of roof mounted PV modules on the fire ratings of roofs—a minimum separation gap and a sheet metal flashing to block the passage of flames between the PV module and the roof assembly.

Considerations of Module Position on Roof Deck During Spread of Flame Tests, July 24, 2012.

The fifth project included a series of experiments to investigate a modification of the current UL 1703 spread of flame test to (1) expose a PV module to flames originating from the UL 790 (ASTM E108) ignition source, (2) allow those flames to generate on a representative roof section, and (3) observe the propagation of the flames underneath the candidate PV module being tested. The repositioning of the PV module was conducted to investigate an application of the first item (roof)/second item (module) ignition sequence. This concept was investigated to refine the understanding of the effect of a rack mounted PV array on the fire rating of a Class A roof.

Building Integrated Photovoltaics

The work described above does not apply to Building Integrated Photovoltaic (BIPV) installations. Since BIPV become the roof, they must comply with the fire classification requirements for roof assemblies as described in UL Standard 790.

International Building Code

The 2012 International Building Code includes the following requirement: "1509.6.2 Fire Classification. Rooftop mounted photovoltaic systems shall have the same fire classification as the roof assembly as defined required by Section 1505."

A Solar ABCs White Paper, Impacts on Photovoltaic Installations of Changes to the 2012 International Codes, discusses the fire classification change and other code changes affective photovoltaic installations. The development of the 2015 Edition of the model codes developed by the International Code Council is underway. Code proposals were accepted to change the language in the 2015 International Building Code. There is also an opportunity to submit Public Comments on the actions taken at the Code Development Hearings by the Code Development Committees, and those Public Comments will be heard at the Final Action Hearings on October 24-28, 2012 in Portland, Oregon.

Representative Response/Potential Mitigation Strategies

The following material is reprinted from Solar Panel Fires and Electrical Hazards—InterNACHI http://www.nachi.org/solar-panel-fire-electrical-hazards.ht m#ixzz27InjsOlR *(accessed 7/24/2012)*

Installed properly, PV solar panels do not cause fires. Most PV modules are tested by Underwriters Laboratories (UL), which subjects them to the rigors of everyday use before they are certified. In the rare cases where PV modules have been implicated in house fires, the cause has been electrical arcing due to improper installation, faulty wiring or insufficient insulation http://www.nachi.org/ solar-panel-fire-electrical-hazards.html.

PV Systems and House Fires

PV systems may be a hazard in the case of a house fire, particularly if firefighters are not aware that a system is installed. Some of these hazards are as follows:

- The conduit leading from PV panels to an inverter may remain live with direct current even after the main service panel has been shut off. Firefighters who unknowingly sever live lines are vulnerable to electrical shock. Some firefighters carry a "hot stick" that aids them in finding live wires, but it does not detect direct current.
- Solar panels and batteries contain toxic chemicals that may be released in a fire and are dangerous if inhaled.
- PV modules may become slippery and pose a slip-and-fall risk to inspectors, technicians and firefighters.
- Solar panels may block key points and pathways that inspectors, technicians and firefighters would otherwise use to mount, navigate and dismount from a roof.
- PV modules may inhibit ventilation of a fire in prime roof locations.
- The added weight of a solar panel array may lead to roof collapse if the integrity of the structure is already compromised by fire.

InterNACHI inspectors may want to check for the following design elements that will prevent PV modules from exacerbating the dangers of a house fire:

- Photovoltaic systems should be installed and subsequently inspected regularly by a qualified professional.
- PV systems should be labeled in a clear and systematic manner to ensure that technicians and firefighters can quickly and easily identify key elements of the system. The main service disconnect panel should be clearly labeled on the outside cover, if it is operable from the outside without opening. Both interior and exterior portions of live conduit should be labeled every 10 ft. Batteries should also be clearly labeled.
- A rooftop shutoff valve should be present. This switch could be utilized to disable the direct current running from the solar panels through the conduit.
- The roof should have sufficient pathways and perimeter space around the PV modules so that inspectors and firefighters can traverse the roof safely.
- There should be a section of the roof left vacant so that it may be ventilated, if necessary.
- Check for damage from rodents and other pests, which could compromise wiring or insulation.
- There should be an integrated arc-fault detection device present in the solar panels, which shuts down individual panels in the case of a malfunction, such as arcing.
- During the permitting process when the PV system is installed, the local fire department should be given a set of the plans to refer to in case of emergency.

Lightweight Engineered Lumber
Issue
Concerns with lightweight engineered lumber (LEL) include decreased thermal resistance to fire and contribution to fuel load. Studies, including those at UL and NRC Canada have highlighted these issues.

UL Test Program

The study, *Structural Stability of Engineered Lumber in Fire Conditions*, involved numerous tests, and a number of project reports and a summary video for firefighters are available on the UL website http://www.ul.com/global/eng/pages/offerings/industries/buildingmaterials/fire/fireservice/lightweight/ (accessed 6/29/2012). The video includes the following summary points from the tests:

- Lightweight assemblies, whether protection or non-protected, fail significantly faster than legacy assemblies.
- Legacy assemblies tend to fail locally while lightweight assemblies tend to fail globally.
- Preheating of wood structural members cause weakening of the structure prior to direct fire involvement.
  - This preheating has a greater impact on the performance of light weight assemblies because of their reduced mass, use of composite materials, and reliance on multiple connectors.

The tests have also been reported on by NFPA, which notes, "Methodologically, the (UL) study aimed to provide 'apples to apples' comparisons among assemblies and to show how different construction materials, including traditional lumber, fared in different types of fires... The experiments documented striking differences between traditional and engineered systems. For example, a traditionally constructed floor system, without a drywall ceiling to protect its underside, withstood the test fire for 18 min. By comparison, a similar system using engineered wooden I-beams survived for about six minutes" http://www.nfpa.org/publicJournalDetail.asp?categoryID=1857&itemID=43878&src=NFPA Journal&cookie%5Ftest=1 (accessed 6/29/2012).

NRC Canada

The National Research Council (NRC) Canada also has been studying this issue. The following summary is also taken from the NFPA website http://www.nfpa.org/publicJournalDetail.asp?categoryID=1857&itemID=43878&src=NFPAJournal&cookie%5Ftest=1. (accessed 6/29/2012)

The NRC study, *Fire Performance of Houses: Phase I Study of Unprotected Floor Assemblies in Basement Fire Scenarios*, was also released last December and sought to establish the "typical sequence of events such as the smoke alarm activation, onset of untenable conditions, and structural failure of test assemblies" in a simulated two-story, single-family house with a basement. "With the relatively severe fire scenarios used in the experiments, the times to reach structural failure for the wood I-joist, steel C-joist, metal plate, and metal web wood truss assemblies were 35 to 60 % shorter than that for the [traditional] solid wood joist assembly," the study reported. In every instance, the floors failed, "characterized by a sharp increase in floor deflection and usually accompanied by heavy flame penetration through the test assemblies, as well as by a sharp increase in compartment temperature above the test floor assemblies."

The report can be found directly at http://www.isfsi.org/uploads/Part5.pdf. A full set of NRC publications related to fire and building performance can be found at http://www.nrc-cnrc.gc.ca/eng/publications/index.html

Double Skinned Façade/Cavity Walls

Issue

Using poor thermally insulated materials would give higher heat lost, hence increasing the cooling or heating load of the heating, ventilation and air-conditioning system. New architectural features such as double-skin façade might give a lower heat lost rate. However, as discussed in above, it is much easier to onset flashover for buildings with a thermally insulated façade. Heat generated in a fire would be trapped to give rapid rise of room air temperature. It appears that poor thermally insulated materials like glass with higher heat lost might be safer in a fire. (Chow, C. L. and Chow, W. K., "Fire Safety Concern on Well-Sealed Green Buildings with Low OTTVs" (2010). *International High Performance Buildings Conference*. Paper 1. http://docs.lib.purdue.edu/ihpbc/1 (accessed 7/24/2012)

Representative Response/Potential Mitigation Strategies

Experimental study on smoke movement leading to glass damages in double-skinned façade (Chow et al. 2007)

The fire hazard of the new architectural feature—double-skinned façade was examined experimentally. Full-scale burning tests on part of the design feature were carried out in a facility developed in a remote area in Northeast China. A total of eight tests were performed to demonstrate how the depth of cavity of a double-skinned façade affects the smoke movement. Surface temperature and heat flux received on the test panels are presented. Cracking patterns found on the glass panels are also observed. The measured results would give the possible smoke movement pattern inside the air cavity. By examining the results for cavity depth of 0.5, 1.0 and 1.5 m, it is found that a deeper cavity might give better safety under the scenario studied. The outer glass panel would be broken rapidly for the cavity of 0.5 m deep. Double-skinned façade with a cavity of 1.0 m deep appeared to be very risky as glass panels above broke most among the different cavity depths. The inner glass panel might be broken before the outer panel. This would give an undesirable outcome. Other separation distances of the two skins should be further examined to give optimum design of cavity depth. Other factors affecting flame and smoke movement should be further investigated.

Ding, W., Hasemi, Y. and Yamada, T., 2005. Smoke Control Using A Double-skin Facade. *Fire Safety Science* 8: 1327–1337. doi:10.3801/IAFSS.FSS.8-1327

Usually for a building with a multistory double-skin façade, smoke of a fire room escaping through the inner façade into the intermediate space between the two skins may accumulate and spread horizontally and/or vertically to other rooms that have openings connected to the intermediate space for the purpose of natural ventilation. However if smoke pressure in the intermediate space can be kept lower than that of the room, smoke spread through the openings will be prevented. Considering similarity of smoke movement and stack natural ventilation, in this paper a double-skin façade used for natural ventilation is also considered for smoke control. As the results, it is proved that smoke spread can be prevented with suitable opening arrangements. Therefore natural ventilation and smoke control can be realized through one system.

Structural Insulated Panels (SIP)

Issue

A structural insulated panel (SIP) consists of two layers of facing material bonded to a low density insulating material in the middle. The face materials can be metal, cement, gypsum board, wood or oriented strand board (OSB). While some studies have been conducted of SIPs usedin residential installations, such as the UK test program listed below, several fires have occurred recently in high-rise buildings with metal clad structural insulating panels.

The performance in fire of structural insulated panels (BD2710)—http://www. communities.gov.uk/documents/planningandbuilding/pdf/1798045.pdf   (accessed 7/24/12)

"A structural insulated panel (SIP) consists of two high density face layers bonded both sides to a low density, cellular core substrate. The structural bond between the layers is essential in providing the required load bearing capacity of the panel. The face layers may be cement or gypsum based boards or wood based boards such as oriented strand board (OSB). The materials used for the core substrate range from synthetic rigid foam cores such as extruded or expanded polystyrene, polyurethane, polyisocyanurate to inorganic mineral fibers. The project has identified collapse of the floor as the predominant mode of failure of the building systems tested as part of this work program based on fire penetration into the floor/ceiling void and combustion of the oriented strand board webs of the engineered floor joists leading to loss of load bearing capacity and runaway deflection."

Recently, fires in the UAE, Shanghai and South Korea have raised concerns with façade of metal SIP with potentially non-fire rated insulation. For example:

Hundreds of skyscrapers across the UAE are wrapped in dangerous non fire-rated aluminium cladding panels that may put lives in danger in the event of a fire, Gulf News has learnt. A top executive …, who spoke on condition of anonymity, (stated) "At least 500 towers in the country have non fire-rated panels installed over the last 25 years" http://gulfnews.com/news/gulf/uae/housing-property/tower-cladding-in-uae-fuels-fire-1.1016836 (accessed 7/24/12).

While some of this may be a result of general construction practice, these systems are used in building retrofit, including energy retrofit. In Busan, South Korea, while the cladding did play a role, the insulating material seems unclear, with some indications that the exterior finish (paint) was to blame. "It took just 20 min for the blaze that started at a trash collection site on the fourth floor to travel up to the 38th floor. The building's concrete body was covered with aluminum panels for aesthetic effect, filled with glass fiber for insulation and coated with flammable paint causing the flames to spread upward quickly." [http://english.chosun.com/site/data/html_dir/20 10/10/04/2010100401225.html (last accessed 7/24/2012)] "Experts said yesterday's fire was their worst nightmare come true. They said high-rises around the nation are particularly vulnerable to fire because they are built with flammable exterior materials. "The sprinkler system worked," said Lee Gap-jin, a senior official at Busan Fire Department's Geumgang branch. But building codes didn't require sprinklers on the janitors' room floor. "And the flames on the exterior walls were too strong, so the sprinklers weren't much help," he said. (http://koreabridge.net/post/haeundae-highrise-fire-busan-marine-city-burns, last accessed 7/24/12).

In Shanghai, the fire was reported as welders igniting combustible scaffolding, which then ignited non-fire rated insulation. "The rigid polyurethane foam pasted on the surface of the other two buildings, which some experts suspect turned the fire into a disaster, will be replaced by fire-resistant materials," Zhang Renliang, head of the Jing'an district government, said at a press conference held regarding the 2012 fire which cost 58 lives (http://www.china.org.cn/china/2010-11/24/content_21407451.htm, last accessed 7/24/12).

http://www.koreaherald.com/national/Detail.jsp? newsMLId=20101001000621 last accessed 7/24/12

http://www.bbc.co.uk/news/world-asia-pacific-11760467 last accessed 7/24/12

Representative Response/Potential Mitigation Strategies

A first response is to use only approved/listed product. However, since it is not clear that these systems are performing as expected, further investigation into fire performance and testing may be warranted.

Vegetative Roof Systems
Issue
Issues with vegetative roof systems include flammability of materials, flammability of vegetation (fire spread), and firefighter access issues (see NASFM 2010). From the flammability of material side, however, steps have been taken to produce test standards to help offset this issue:

Representative Response/Potential Mitigation Strategies

Approval Standard for Vegetative Roof Systems—FM Approvals—contains general information, general requirements, performance requirements, and operations requirements—available from: http://www.fmglobal.com/assets/pdf/fmapprovals/4 477.pdf, last accessed 7/24/12.

ANSI/SPRI VF-1 External Fire Design Standard for Vegetative Roofs, available from: http://www.greenroofs.org/resources/ANSI_SPRI_VF_1_Extrernal_Fire_ Design_Standard_for_Vegetative_Roofs_Jan_2010.pdf, last accessed 7/24/12.

Flammability and Toxicity of Foam Insulations
Issue
Several issues with the increasing use of insulating materials for increased energy efficiency were identified. These include hazards associated with do-it-yourself installation of foil and other insulation systems, leading to shock hazards and ignition hazards (e.g., see Australia Home Insulation Program issues), concerns that additional insulation might lead to higher compartment fire temperatures, and that additional insulation adds more fuel load. There have also been issues raised with respect to flame retardants in foam insulations causing health hazards (e.g., see http://media.apps.chicagotribune.com/flames/index.html, last accessed 9/24/2012).

Representative Response/Potential Mitigation Strategies

Potential mitigation options for flammability issues exist with respect to thermal barrier requirements in the US, at least (IBC and IRC). In fact, some argue that the presence of the thermal barrier means flame retardants can be removed. However, other resources note that flame retardants are still needed given the overall flammability and burning characteristics of the foam insulation (e.g., see Hirschler 2008, *Polymers for Advanced Technologies* and Ohlemiller and Shields 2008, *NIST Technical Note 1493*). In the end, the efficacy of the thermal barrier only exists if the thermal barrier is in place and functioning properly when needed, and if not, hazards remain. This area seems to warrant further study.

Hirschler, M.M., "Polyurethane foam and fire safety," *Polym. Adv. Technol.* 2008; 19: 521–529 (2008).

Ohlemiller, T.J. and Shields, J.R., Aspects of the Fire Behavior of Thermoplastic Materials, NIST Technical Note 1493, Gaithersburg, MD (2008).

Appendix C
International Survey and Responses

B. Meacham et al., *Fire Safety Challenges of Green Buildings*, SpringerBriefs in Fire, DOI: 10.1007/978-1-4614-8142-3, © Fire Protection Research Foundation 2012

Appendix C

Country/entity	Fire incident experience/tracking Fire incident experience/tracking involving green building elements in green buildings	Risk-based assessment of green building elements
Australia		
New South Wales Fire Brigade	The structures that subscribe to the National Built Environment Rating System (NABERS) are usually commercial or government buildings. In most cases they are relatively new and range from modern high rise premises in the city (eg. No. 1 Bligh St.) to restored and renovated federation style buildings (eg. 39 Hunter St.) The building codes also provide for prescribed or engineered fire safety solutions. There are no specific AIRS codes for "Green" buildings therefore it is very difficult to determine if there have been any fires or dominant fire causes in these buildings	*Ceiling Insulation*: FIRU have experienced major concerns with this issue particularly in residential, nursing homes and aged care facilities. Cellulose fibre insulation in close proximity to downlights and insulation batts including non-compliance with electrical wiring rules have been the dominant concerns. AIRS analysis for insulation fires 29/02/2008 to 22/06/2011. Data provided by SIS: The data includes 102 incidents that occurred in metropolitan, regional and country areas. Of these incidents 75 were directly related to downlights and their associated transformers in close proximity to ceiling insulation. Some of the fires resulted in substantial property damage. FIRU have reports of residential structures constructed of insulated sandwich panels in locations ranging from Broken Hill to Thredbo. *Laminated Timber I-Beams*: No specific AIRS codes for Laminated Timber I-Beams. A combination of open plan living and modern furnishings (eg. polyurethane foam settees, etc.) can create fuel packages that will reach temperatures of 1,000–1,200 °C that can rapidly weaken structural elements. *Photovoltaic Solar Installations*: No specific AIRS codes for PV solar installations. It is estimated that NSW has about 150,000 PV solar installations. Predominant causes include faulty components and incorrect installation. (2010 to August 2012 = nine reported incidents) FIRU research indicates major problems with PV solar installations. No remote solar isolation switching, no DC rest device for Firefighters and no 24/7 availability of qualified PV solar electricians. This has been reflected in reports from around Australia, Germany and the US. From FIRU research it appears that major fire services throughout Australasia, United Kingdom, Germany and the United States have experienced concerns with "Green" building elements. When attending fire calls operational Firefighters are unable to readily determine if it is a "Green" building or one with "Green" building elements	FRNSW has a Risk Based Approach for all incidents, including all types of structures. This is supported by a broad range of Standard Operational Guidelines (SOGs), Safety Bulletins and Operations Bulletins. No specific effort related to green buildings.

(continued)

Appendix C (continued)

Country/entity	Fire incident experience/tracking in green buildings	Fire incident experience/tracking involving green building elements	Risk-based assessment of green building elements
Japan			
National Institute of Land and Infrastructure Management	I do not have any information. Fire department and fire and disaster management agency may have fire data, but they will not supply data by the duty to protect privileged information	The Japan Association for Fire Science and Engineering annual symposium was held on May 21–22, 2012. There were two paper related to green building elements as attached files. (1) Hiroyuki Tamura (National Research Institute of fire and disaster) et al., Electricity generation characteristic of a photovoltaic module in a fire. (2) Sanae Matsushima (National Research Institute of fire and disaster) et al., Electricity generation characteristic of fire-damaged photovoltaic module. (in Japanese)	See papers (only known activity).
Netherlands			
TNO (Technical Research Organization Netherlands)	No information is available within the Netherlands about examples of fire incidents in case of green buildings, because this information is not taken into account within the existing databases	No information is available within the Netherlands about examples of fire incidents in case of green buildings, because this information is not taken into account within the existing databases	I also see that no research has been started on this topic within the Netherlands up till now
New Zealand			
Department of Building and Housing	We're not aware of any fires in 'green buildings' in New Zealand	However, there have certainly been instances of fires involving building materials that are designed to provide energy efficiency. There have been instances of insulation overheating due to contact with downlights etc. I don't have any specific instances but New Zealand Fire Service may be able to provide statistics if you'd like us to follow up.' We have also had cases of stapling through electrical wiring when retrofitting sub floor insulation—not starting fires but obviously electrocution is an issue	Not aware of any activities

(continued)

Appendix C (continued)

Country/entity	Fire incident experience/tracking Fire incident experience/tracking involving green building elements in green buildings	Risk-based assessment of green building elements
BRANZ	No specific data. The New Zealand Fire Service might have some.	Scoping study related to building sustainability and fire safety design interactions might be helpful—http://www.branz.co.nz/cms_show_download.php?id=71673351 5027fe462618888 1 f674635d51e3cfb0	
Norway		
Norwegian Building Authority	I have no information on this subject, but I have forwarded your questions to The Norwegian Insurance Approval Board (responsible of fire statistics collected by the insurance companies) and The Norwegian Fire Protection Association. (BJM note: no response from either)	I have no information on this subject, but I have forwarded your questions to The Norwegian Insurance Approval Board (responsible of fire statistics collected by the insurance companies) and The Norwegian Fire Protection Association. (BJM note: no response from either)	
Scotland		

(continued)

Appendix C (continued)

Country/entity	Fire incident experience/tracking in green buildings	Fire incident experience/tracking involving green building elements	Risk-based assessment of green building elements
Scottish Government: Building Standards Division	It might also be worth pointing out to Brian that the National fire statistics do not identify fires in buildings with green credentials and/or green certification. However, BRE due carry out fire investigations on behalf of CLG and may well have identified green issues as being a contributory factor to the fire development and subsequent damage. Contact Martin Shipp from BRE at ShippM@bre.co.uk. Martin should also be able to provide contact details of other fire investigators from other research institutions/laboratories, the insurance industry and fire and rescue services who will all have experience in these issues	The only research we have sponsored through CLG was in relation to the revision of BR 187 'External fire spread: building separation and boundary distances.' BRE intend to publish this document in the Autumn but many questions remain unanswered for highly insulated buildings and BRE have suggested 'further' research in this field. Having said that, BRE have published a very good scoping study on the type of issues (http://www.communities.gov.uk/documents/planningandbuilding/pdf/1795639.pdf). Feel free to forward this to Brian as it is in the public domain anyway	Finally, BSD have been involved with a UK wide steering group looking at developing a codified methodology to hazard classification of buildings. The intention is to improve fire-fighter safety when carrying out fire-fighting and rescue operations. This could feed into the risk based approach that Brian touches on in his email Contact Dave Berry at dave.berry@farmss.co.uk
Spain			
Eduardo Torroja Institute for Construction Science	No such incident data collected	No such incident data collected	No research activities in this area

(continued)

Appendix C (continued)

Country/entity	Fire incident experience/tracking in green buildings	Fire incident experience/tracking involving green building elements	Risk-based assessment of green building elements
Sweden			
Boverket (National Board for Building and Housing)	Boverket haven't got any information on fires in green buildings. The number of buildings built with some sort of certification (LEED etc.) is rather small in Sweden and the statistics from fire incidents does not take into account if it is a green building or not. Responsible for the statistics is the Swedish Civil Contingencies Agency (MSB). You can contact them if you want Swedish fire statistics in general, www.msb.se. Our contact in MSB, Malin, can help you find the right person *malin.p ettersson@msb.se*	One area that I know has been highlighted is risk to the environment and fire risks in connection with recycling. Growing risk areas are fires in recycling stations close to a building or in buildings (mainly dwellings) but also risk of leakage of contaminated water after putting out a fire in recycling sites	I don't know of any special research project regarding green buildings and fire risk in Sweden. The Swedish Fire Research Board (Brandforsk) might know of some on-going projects. http://www.brandforsk.se/eng. You could also contact the research institutes directly. As you might know Lund University (www.brand.lth.se) and SP (Technical Research Institute of Sweden, www.sp.se) are the two main fire research institutes in Sweden
USA			
National Fire Protection Association	No such incident data collected	No such incident data collected	No specific research activities in this area, other than FPRF effort, identified

Appendix D
Detailed Matrices of Green Elements and Potential Fire Hazards

D.1 Structural Materials and Systems

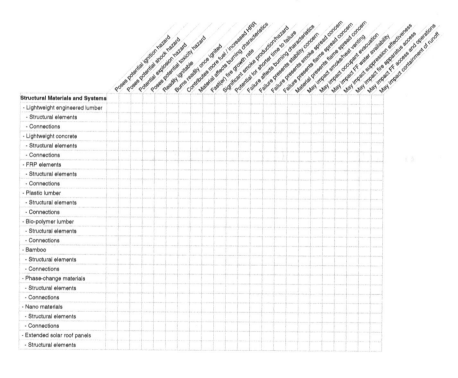

The matrix lists the following column headers (rotated):

- Poses potential ignition hazard
- Poses potential shock hazard
- Potential explosion hazard
- Poses potential toxicity hazard
- Readily ignitable
- Burns readily once ignited
- Contributes more fuel / increased HRR
- Material affects burning characteristics
- Fast(er) fire growth rate
- Significant smoke production/hazard
- Potential for shorter time to failure
- Failure affects burning characteristics
- Failure presents stability concern
- Failure presents smoke spread concern
- Material presents flame spread concern
- May impact smoke/heat venting
- May impact occupant evacuation
- May impact FF water availability
- May impact fire suppression effectiveness
- May impact fire apparatus access
- May impact FF access and operations
- May impact containment of runoff

Structural Materials and Systems
- Lightweight engineered lumber
  - Structural elements
  - Connections
- Lightweight concrete
  - Structural elements
  - Connections
- FRP elements
  - Structural elements
  - Connections
- Plastic lumber
  - Structural elements
  - Connections
- Bio-polymer lumber
  - Structural elements
  - Connections
- Bamboo
  - Structural elements
  - Connections
- Phase-change materials
  - Structural elements
  - Connections
- Nano materials
  - Structural elements
  - Connections
- Extended solar roof panels
  - Structural elements

B. Meacham et al., *Fire Safety Challenges of Green Buildings*, SpringerBriefs in Fire,
DOI: 10.1007/978-1-4614-8142-3, © Fire Protection Research Foundation 2012

D.2 Exterior Materials and Systems

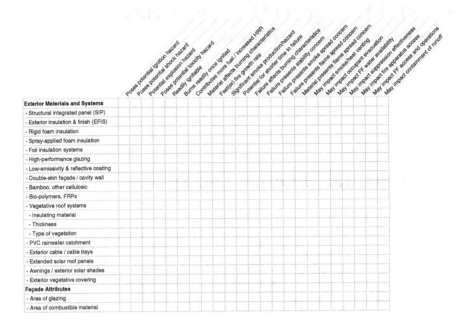

D.3 Interior Materials and Finishes

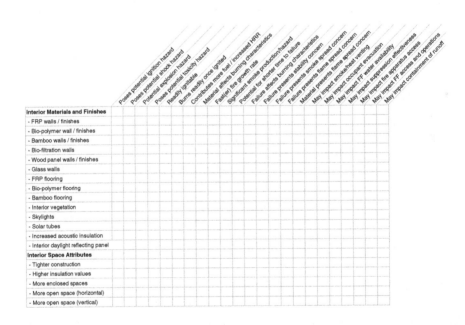

D.4 Building Systems and Issues

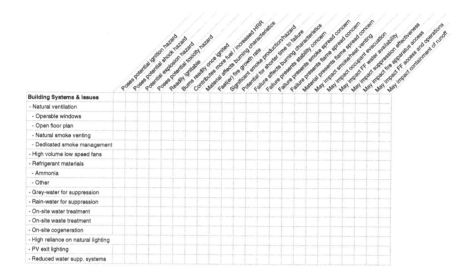

D.5 Alternative Energy Systems

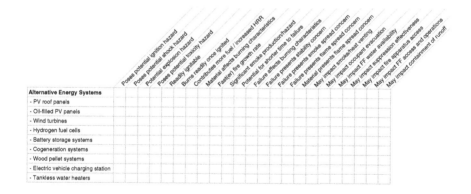

D.6 Site Issues

Site Issues	Poses potential ignition hazard	Poses potential shock hazard	Potential explosion hazard	Poses potential toxicity hazard	Readily ignitable	Burns readily once ignited	Contributes more fuel / increased HRR	Material affects burning characteristics	Fast(er) fire growth rate	Significant smoke production/hazard	Potential for shorter time to failure	Failure affects burning characteristics	Failure presents stability concern	Failure presents smoke spread concern	Material presents flame spread concern	May impact smoke/heat venting	May impact occupant evacuation	May impact FF water availability	May impact suppression effectiveness	May impact fire apparatus access	May impact FF access and operations	May impact containment of runoff
- Permeable concrete systems																						
- Permeables asphalt paving																						
- Use of pavers																						
- Extent (area) of lawn																						
- Water catchment / features																						
- Vegetation for shading																						
- Building orientation																						
- Increased building density																						
- Localized energy production																						
- Localized water treatment																						
- Localized waste treatment																						
- Reduced water supply																						
- Hydrogen infrastructure																						
- Community charging stations																						

Appendix E
Illustration of Relative Hazard Ranking with Detailed Matrix

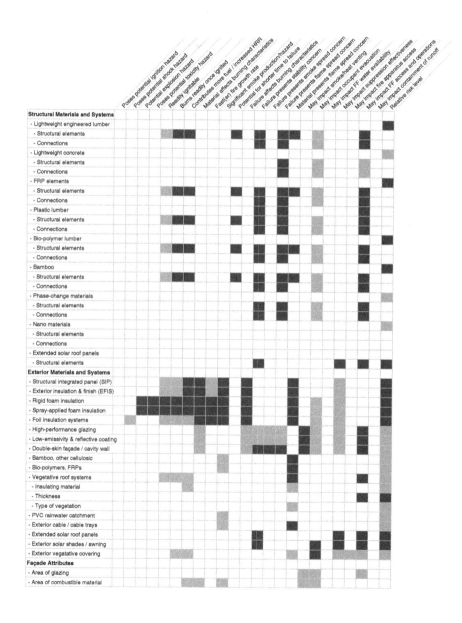

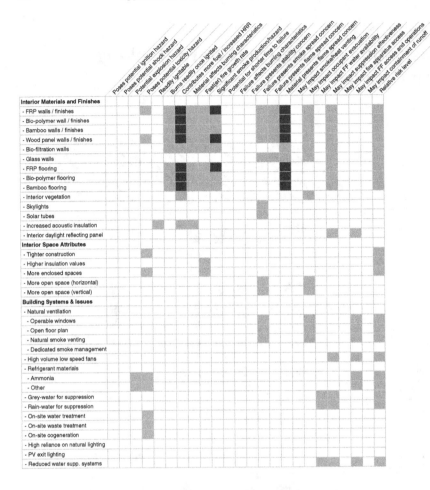

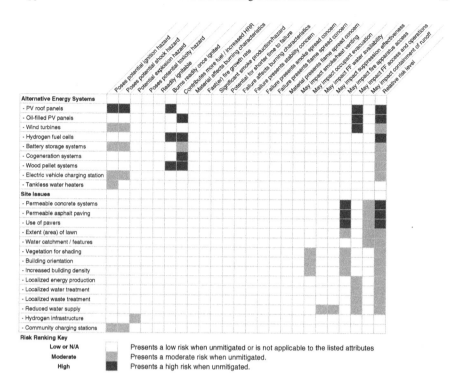

Risk Ranking Key

Low or N/A	Presents a low risk when unmitigated or is not applicable to the listed attributes
Moderate	Presents a moderate risk when unmitigated.
High	Presents a high risk when unmitigated.

Appendix F
Tabular Representation of Fire Hazards with Green Building Elements

Appendix F

Material/system/attribute	Hazard	Concern level	Potential mitigation strategies
Structural materials and systems			
• Lightweight engineered lumber	Can fail more quickly. Contributes to fuel load. Impact for egress and FF. Stability issues	High	Fire resistive barrier (e.g., fire rated gypsum). Approved/listed materials. Sprinklers
• Lightweight concrete	Can spall more explosively if not treated with fiber. Can fail more quickly. FF and stability issues	Moderate	Require fibers for strength. Approved/listed materials
• FRP elements	Can fail more quickly. Contributes to fuel load. Impact for egress and FF. Stability issues	High	Require formulations with high ignition temperatures, low flame spread and low smoke production; cover with thermal barrier or intumescing cover. Approved/listed materials
• Plastic lumber	Can fail more quickly. Contributes to fuel load. Impact for egress and FF. Stability issues	High	Fire resistive barrier (e.g., fire rated gypsum). Approved/listed materials. Sprinklers
• Bio-polymer lumber	Can fail more quickly. Contributes to fuel load. Impact for egress and FF. Stability issues	High	Fire resistive barrier (e.g., fire rated gypsum). Approved/listed materials. Sprinklers
• Bamboo	Can fail more quickly. Contributes to fuel load. Impact for egress and FF. Stability issues	Moderate	Fire resistive barrier (e.g., fire rated gypsum). Approved/listed materials. Sprinklers

(continued)

B. Meacham et al., *Fire Safety Challenges of Green Buildings*, SpringerBriefs in Fire, DOI: 10.1007/978-1-4614-8142-3, © Fire Protection Research Foundation 2012

Appendix F (continued)

Material/system/ attribute	Hazard	Concern level	Potential mitigation strategies
• Phase-change materials	Unknown	Unknown	Research and testing. Approved/listed materials
• Nano materials	Unknown	Unknown	Research and testing. Approved/listed materials
• Extended solar roof panels	Can create hazard to FF if fails. Impacts FF access	Moderate	Provide fire proofing. Assure options for FF access. Approved/ listed materials
Exterior materials and systems			
• Structural integrated panel (SIP)	If fail, insulation can contribute to flame spread, smoke production and fuel load	High	Approved/listed materials. Assure proper seal ing of panels. Take care during installation, including retrofits, relative to potential sources of ignition
• Exterior insulation and finish (EFIS)	If fail, insulation can contribute to flame spread, smoke production and fuel load	High	Approved/listed materials. Assure proper seal ing of panels. Take care during installation, including retrofits, relative to potential sources of ignition
• Rigid foam insulation	Can contribute to flame spread, smoke and toxic product development and fuel load	High	Fire resistive barrier (e.g., fire rated gypsum). Approved/listed materials. Flame retardants. Sprinklers
• Spray-applied foam insulation	Can contribute to flame spread, smoke and toxic product development and fuel load	High	Fire resistive barrier (e.g., fire rated gypsum). Approved/listed materials. Flame retardants. Sprinklers
• Foil insulation systems	Can contribute to shock hazard for installers. Can contribute to flame spread and fuel load	High	Fire resistive barrier (e.g., fire rated gypsum). Approved/listed materials. Sprinklers
• High-performance glazing	Can change thermal characteristics of compartment for burning. Can impact FF access	Moderate	Sprinklers. Assure adequate FD access. Assure mechanism for FD smoke/heat venting. Approved/ listed materials

(continued)

Appendix F (continued)

Material/system/ attribute	Hazard	Concern level	Potential mitigation strategies
• Low-emissivity and reflective coating	Can change thermal characteristics of compartment for burning. Can impact FF access	Moderate	Sprinklers. Assure adequate FD access. Assure mechanism for FD smoke/heat venting. Approved/ listed materials
• Double-skin facade	Can change thermal characteristics of compartment for burning. Can impact FF access. Can present 'chimney' for vertical smoke and flame spread if not properly fire stopped	Moderate	Appropriate fire stop between floors. Sprinklers may have some benefit (sprinklered building). Assure mechanism for FD smoke/heat venting. Approved/ listed materials
• Bamboo, other cellulosic	Can contribute to flame spread, smoke development and fuel load	Moderate	Approved/listed materials. Flame retardant treatments. Sprinklers
• Bio-polymers, FRPs	Can contribute to flame spread, smoke development and fuel load	Low	Approved/listed materials. Flame retardant treatments. Sprinklers
• Vegetative roof systems	Can contribute to fire load, spread of fire, impact FF operations, impact smoke and heat venting, contribute to stability issues	Moderate	Manage fire risk of vegetation. Assure use of fire tested components. Provide adequate area for FD access, smoke/heat venting, and other operations. Approved/ listed materials
• PVC rainwater catchment	Can contribute additional fuel load	Low	Limit volume
• Exterior cable/cable trays	Can contribute additional fuel load	Low	Limit volume. Approved/ listed materials
Facade Attributes			
• Area of glazing	Can present more opportunity for breakage and subsequent fire spread and/or barrier to FF access depending on type	Moderate	
• Area of combustible material	Larger area (volume) provides increased fuel load	High	Limit volume
• Awnings	Impacts FF access	Low	
• Exterior vegetative covering	Can impact FF access and present WUI issue	Low	Limit volume

(continued)

Appendix F (continued)

Material/system/ attribute	Hazard	Concern level	Potential mitigation strategies
Interior materials and finishes			
• FRP walls/ finishes	Can contribute to flame spread, smoke development and fuel load	Moderate	Approved/listed materials. Flame retardant treatments. Sprinklers
• Bio-polymer wall/ finishes	Can contribute to flame spread, smoke development and fuel load	Moderate	Approved/listed materials. Flame retardant treatments. Sprinklers
• Bamboo walls/ finishes	Can contribute to flame spread, smoke development and fuel load	Moderate	Approved/listed materials. Flame retardant treatments. Sprinklers
• Wood panel walls/ finishes	Can contribute to flame spread, smoke development and fuel load	Moderate	Approved/listed materials. Flame retardant treatments. Sprinklers
• Bio-filtration walls	Can contribute to flame spread, smoke spread a nd fuel load	Low	Approved/listed materials
• Glass walls	May not provide adequate fire barrier alone	Moderate	Approved/listed materials. Sprinklers
• FRP flooring	Can contribute to flame spread, smoke development and fuel load	Low	Approved/listed materials. Flame retardant treatments. Sprinklers
• Bio-polymer flooring	Can contribute to flame spread, smoke development and fuel load	Low	Approved/listed materials. Flame retardant treatments. Sprinklers
• Bamboo flooring	Can contribute to flame spread, smoke development and fuel load	Low	Approved/listed materials. Flame retardant treatments. Sprinklers

(continued)

Appendix F (continued)

Material/system/ attribute	Hazard	Concern level	Potential mitigation strategies
Interior Space Attributes			
• Tighter construction	Can change burning characteristics of compartments. Can result in negative heal th effects, moisture and related issues	Moderate	Assure adequate air changes and filtering. Approved/ listed materials
• Higher insulation values	Can change compart- ment burning characteristics, result in additional fuel load and lead to impacts to FF access	Moderate	Approved/listed materials. Sprinklers
• More enclosed spaces	Can result in challenges in finding fire source	Low	Sprinklers
• More open space (horizontal)	Can contribute to fire and smoke spread	Moderate	Sprinklers
• More open space (vertical)	Can contribute to fire and smoke spread	Moderate	Sprinklers
• Interior vegetation	Can contribute fuel load. Can impact FF operations	Low	Sprinklers
• Skylights	Can contribute to fire and smoke spread	Low	Approved/listed materials. Sprinklers
• Solar tubes	Can contribute to fire and smoke spread	Low	Approved/listed materials. Sprinklers
• Increased acoustic insulation	Can change compart- ment burning characteristics, result in additional fuel load and lead to impacts to FF access	Moderate	Approved/listed materials. Sprinklers
Building systems and issues			
• Natural ventilation	Can impact ability to control smoke. Can influence smoke movement depend- ing on environmen- tal conditions	Moderate	Dedicated smoke management system. Sprinklers. Dedicated FF smoke venting
• High volume low speed fans	Can influence sprin- kler and detector performance	Moderate	Additional sprinkler protection beyond code requirements

(continued)

Appendix F (continued)

Material/system/ attribute	Hazard	Concern level	Potential mitigation strategies
• Refrigerant materials	Can provide different burning, toxic- ity, and HazMat concerns	Moderate	Approved/listed materials. Treat and protect appropriate to material hazards
• Grey-water for suppression	Can have impact of water availability for suppression. Could have impact on MIC issues with sprinkler and hydrant systems	Low	Assure water is prop- erly treated for use in sprinkler and stand pipe system
• Rain-water for suppression	Can have impact of water availability for suppression. Could have impact on MIC issues with sprinkler and hydrant systems	Low	Assure water is prop- erly treated for use in sprinkler and stand pipe system
• On-site water treatment	Can have impact of water availability for suppression. Could have impact on MIC issues with sprinkler and hydrant systems.	Low	Locate in fire ra ted construction or separate building. Sprinkler
• On-site waste treatment	Can create HazMat and containment issues	Low	Locate in fire ra ted construction or separate building. Sprinkler
• On-site cogeneration	Can present new fire hazards	Low	Locate in fire ra ted construction or separate building. Sprinkler
• High reliance on natural lighting	Can result in larger area of high- performance glazing	Moderate	Consider including of battery powered emergency lighting
• PV exit lighting	Require permanent full lighting to charge material—if used with increased natural lighting may not be effective	Moderate	Consider including of battery powered emergency lighting
• Reduced water supp. systems	Local restrictions or conditions (e.g., drought) may limit water available for suppression	High	Include water storage within building/ on-site to meet mini- mum FP needs

(continued)

Appendix F (continued)

Material/system/ attribute	Hazard	Concern level	Potential mitigation strategies
Alternative energy systems			
• PV roof panels	Presents ignition hazard and contributes to fuel load. Prevents shock hazard to FF. Presents glass breakage hazard	High	Provide thermal barriers between PV cells and combustible roof material. Use noncombustible roof materials. Design roof space for FF access, heat and smoke venting. Have emergency power interruption. Clearly mark. Approved/listed materials
• Oil-filled PV panels	Presents ignition hazard and contributes to fuel load	High	Provide thermal barriers between PV cells and combustible roof material. Use noncombustible roof materials. Design roof space for FF access, heat and smoke venting. Have emergency power interruption. Clearly mark. Approved/ listed materials
• Wind turbines	Potential ignition hazard	Low	Automatic and manual power interruption
• Hydrogen fuel cells	Presents explosion hazard and contributes to fuel load	Moderate	Install in explosion vented or resistant enclosure. Leak detection and automatic shutoff. Clearly mark
• Battery storage systems	Presents ignition hazard and contributes to fuel load. Source of potential shock hazard. My release corrosive or toxic materials if damaged	Low	Provide adequate compartmentation and special suppression. Clearly mark. Approved/listed materials
• Cogeneration systems	Additional fuel load	Low	Provide adequate compartmentation and special suppression. Clearly mark

(continued)

Appendix F (continued)

Material/system/ attribute	Hazard	Concern level	Potential mitigation strategies
• Wood pellet systems	Additional fuel load	Low	Sprinklers
• Electric vehicle charging station	Presents ignition hazard	Low	Adequate shutoffs, shock protection. Clearly mark
• Tankless water heaters	May present ignition hazard	Low	Smoke and CO alarms. Approved/listed materials
Site Issues			
• Permeable concrete systems	May affect pooling of flamable liquid and resulting pool fire, containment, runoff containment issues	Moderate	Appropriate emergency response planning, including spill containment and suppression, and vehicle access
• Permeable asphalt paving	May affect pooling of flamable liquid and resulting pool fire, containment, runoff containment issues	Moderate	Appropriate emergency response planning, including spill containment and suppression, and vehicle access
• Use of pavers	May affect pooling of flamable liquid and resulting pool fire, containment, runoff containment issues. May also have load-carrying issues wrt fire apparatus	Moderate	Appropriate emergency response planning, including spill containment and suppression, and vehicle access
• Extent (area) of lawn	May present fire apparatus access challenges	Low	Appropriate emergency response planning, including vehicle access
• Water catchment/ features	May present fire apparatus access challenges	Low	Appropriate emergency response planning, including vehicle access
• Vegetation for shading	May present fire apparatus access challenges	Low	Appropriate emergency response planning, including vehicle access
• Building orientation	May present fire apparatus access challenges	Low	Appropriate emergency response planning, including vehicle access
• Increased building density	May present fire apparatus access challenges. May increase fire spread potential	Moderate	Appropriate emergency response planning, including vehicle access

(continued)

Appendix F (continued)

Material/system/ attribute	Hazard	Concern level	Potential mitigation strategies
• Localized energy production	May present more challenging fires for FD. May present access issues	Low	Appropriate emergency response planning, including vehicle access
• Localized water treatment	May present more challenging fires for FD. May present access issues. May impact runoff issues (may overload system with runoff)	Low	Appropriate emergency response planning, including vehicle access
• Localized waste treatment	May present more challenging fires for FD. May present access issues. May impact runoff issues	Low	Appropriate emergency response planning, including vehicle access
• Reduced water supply	Local restrictions or conditions (e.g., drought) may limit water available for suppression	High	Appropriate emergency response planning, including vehicle access. Consider local water supply (site or facility)
• Hydrogen infrastructure	May present new and challenging fire and explosion hazards, putting several properties at risk depending on density	Moderate	Appropriate emergency response planning. Appropriate shock protection. Suppression system
• Community charging stations	May present shock hazards for multiple users	Low	Appropriate emergency response planning. Suppression system. Explosion venting/ protection

Appendix G
Assessment of LEED, BREEAM, GREENMARK and the IgCC for Fire Safety Objectives

B. Meacham et al., *Fire Safety Challenges of Green Buildings*, SpringerBriefs in Fire,
DOI: 10.1007/978-1-4614-8142-3, © Fire Protection Research Foundation 2012

References

AS/NZS4630 (2004) Risk management-principles and guidelines. Standards Australia, Sydney, Australia

BRE 2709 (2010) Impact of fire on the environment and building sustainability BD 2709. Department for Communities and Local Government, UK. Available at http://www.communities.gov.uk/documents/planningandbuilding/pdf/1795639.pdf. Accessed 21 Oct 2012

ISO 31000 (2009) Risk management-principles and guidelines. ISO, Geneva

Meacham BJ (2000) Application of a decision-support tool for comparing and ranking risk factors for incorporation into performance-based building regulations. In: Proceedings of the third international conference on performance-based codes and fire safety design methods, SFPE, Bethesda

Meacham BJ (2004) Understanding risk: quantification, perceptions and characterization. J Fire Prot Eng 14(3):199–227

Meacham BJ, Johnson PJ, Charters D, Salisbury M (2008) "Building fire risk analysis", Section 5, Chapter 12, SFPE handbook of fire protection engineering, 4th edn. NFPA and SFPE, Quincy

Meacham BJ, Dembsey NA, Johann M, Schebel K, Tubbs J (2012) Use of small-scale test data to enhance fire-related threat, vulnerability, consequence and risk assessment for passenger rail vehicles. J Homel Secur Emerg Manag 9(1):1–16

NFPA 551, Guide for the evaluation of fire risk assessments. NFPA, Quincy

NAP (1996) Understanding risk: informing decisions in a democratic society. National Academy Press, Washington

Ramachandran G, Charters D (2011) Quantitative risk assessment in fire safety. Spon Press, London

SFPE (2008) SFPE handbook of fire protection engineering. NFPA, Quincy

SFPE (2006) SFPE engineering guide on fire risk assessment. SFPE, Bethesda

Watts J (2008) "Fire risk indexing", Section 5, Chapter 10, SFPE handbook of fire protection engineering, 4th edn. NFPA and SFPE, Quincy